MW01517900

Cellulose Nanocomposites

ACS SYMPOSIUM SERIES **938**

Cellulose Nanocomposites

Processing, Characterization, and Properties

Kristiina Oksman, Editor
Norwegian University of Science and Technology

Mohini Sain, Editor
University of Toronto

Sponsored by the
ACS Division of Cellulose and Renewable Materials

American Chemical Society, Washington, DC

Library of Congress Cataloging-in-Publication Data

American Chemical Society. Meeting (229th : 2005 : San Diego, Calif.)
 Cellulose nanocomposites : processing, characterization, and properties / Kristiina
Oksman, Mohini Saha, editors.

  p. cm.—(ACS symposium series ; 938)

  "Sponsored by the ACS Division of Cellulose and Renewable Materials."

  Includes bibliographical references and index.

  ISBN 13: 978–0–8412–3980–7 (alk. paper)

  ISBN 10: 0–8412–3980–0 (alk. paper)

  1. Nanostructured materials—Congresses. 2. Composite materials—Congresses. 3.
Cellulose—Congresses.

  I. Oksman, Kristiina, 1959- II. Sain, Mohini, 1956- III. American Chemical Society.
Division of Cellulose and Renewable Materials. IV. Title. V. Series.

TA418.9.N35A44   2005
620.1′97—dc22   2006042839

The paper used in this publication meets the minimum requirements of American
National Standard for Information Sciences—Permanence of Paper for Printed Library
Materials, ANSI Z39.48–1984.

PRINTED IN THE UNITED STATES OF AMERICA

Foreword

The ACS Symposium Series was first published in 1974 to provide a mechanism for publishing symposia quickly in book form. The purpose of the series is to publish timely, comprehensive books developed from ACS sponsored symposia based on current scientific research. Occasionally, books are developed from symposia sponsored by other organizations when the topic is of keen interest to the chemistry audience.

Before agreeing to publish a book, the proposed table of contents is reviewed for appropriate and comprehensive coverage and for interest to the audience. Some papers may be excluded to better focus the book; others may be added to provide comprehensiveness. When appropriate, overview or introductory chapters are added. Drafts of chapters are peer-reviewed prior to final acceptance or rejection, and manuscripts are prepared in camera-ready format.

As a rule, only original research papers and original review papers are included in the volumes. Verbatim reproductions of previously published papers are not accepted.

ACS Books Department

Contents

Introduction

Materials Characterization

Nanocomposites Processing and Properties

Indexes

Cellulose Nanocomposites

Introduction

Chapter 1

Introduction to Cellulose Nanocomposites

Mohini Sain[1] and Kristiina Oksman[2]

[1]Centre for Biocomposites and Biomaterials Processing, Faculty
of Forestry, University of Toronto, Toronto, Ontario M5S 3B3, Canada
[2]Department of Engineering Design and Materials, Norwegian University of
Science and Technology, Rich Birkelands vei 2b,
N7491 Trondheim, Norway

This chapter will introduce the reader to the topic of nanocomposites based on cellulose and also give a short description of the content of this book. This is a relatively new research field and there is no other book available on this topic. There is a growing interest on cellulose nanocomposites in developed and developing world and, especially if the nanocomposites are based totally on renewable raw materials. The purpose of this book is to provide information about how to produce nano whiskers or fibrils from different cellulosic sources and the techniques to characterize the nano structures of cellulosic materials and their composite properties. This book will give knowledge of different processing methods for nanocomposites, provide updated information on composites properties and also deal with interesting applications especially in medical field.

Introduction

We all know that sustainable development, reducing pressure on fossil fuel demand and waste reduction are interrelated factors, which could bring significant benefit to the standard of living in the 21st century. It is easy to write these "buzz" words and then to loose our focus on how to realize and deal with the challenges associated with achieving those targets. A recent trend in the modem technology development is to "miniaturize" a system. The word "nano" is now added to the dictionary of materials and manufacturing to encompass the area of nano-materials including nanocomposites. To be honest, not many of these "hi-tech" developments could satisfy the "core" concept of sustainability. One way scientists, researchers and policy makers can address some of the issues related to sustainability is to help develop "miniaturized" materials and manufacturing processes that use renewable resources. Typical examples are the use of agro-and forest resources to manufacture materials that are lighter, smaller and of higher performance. If one follows the biogenesis process for plant and tree growth it is easy to reveal the nature's gift to our modern and miniaturized "nano-world". The backbone of a plant or tree is a polymeric carbohydrate with abundance of a "tiny" structural entity known as "cellulose fibrils". These fibrils are comprised of different hierarchical microstructures commonly known as nano-sized microfibrils with high structural strength and stiffness. These nano sizes fibrils are again combined of a crystalline and amorphous part and this crystalline part is named nanowhisker. The best way to resource these nano-sized microfibrils from nature would be to better understand the biogenetic pathway of woody materials in plants and trees and, then control the biogenesis by stopping the process in a smaller pathway, which allows easy and low energy intensive access of these fibrils in woody plants. Figure 1 demonstrates these different structures from millimeter to nano scale showing wood as an example.

In recent years, scientists and engineers are working together to use the inherent strength and performance of these nano-fibrils and make out of them a new class nano-materials of renewable nature by technologically combining them with natural polymers. This purpose of this book is to combine all recent research and scale up efforts in a text book to provide readers a comprehensive knowledge on an already advanced science and engineering of renewable nano-composite materials, their manufacturing processes and characterization techniques, properties and some potential applications.

Bionanocomposite: A Biological Pathway to Develop Nanomaterials

In the context of this book bionanocomposites are defined as composites made by technological combination of renewable materials, one component of

4

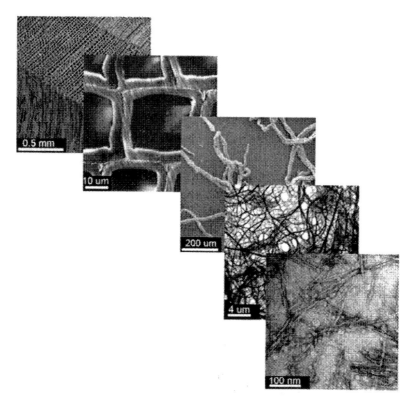

Figure 1. The structure of wood from millimeter to nanometer scale showing the cellular structure of wood, cellulose fibres, microfibres and at last isolated cellulose nano crystals (whiskers).

which has a dimension in the range of 1 to 100 nm to obtain a synergistic performance of the said composites. Although in recent years words such as "Biotechnology", "Nanomaterials" and "Nanotechnology" are very often used to secure investment, government funding and marketing many new products, in essence the objective of this technological development is not always delivered. For example, adding nano-structured clay micro particles in a plastic matrix does not always exploit the nano-structured clay properties and, as a result, in many cases results do not reflect the benefit of a "true" nanomaterial with enhanced performance.

It is anticipated that a fundamental understanding of the reverse engineering of the nature's biogenesis process to generate cellulose fibre bundles from the basic structure of glucose cellulose would provide a platform to generate nano-materials, more specifically bionanofibres from renewable materials such as plants and trees. This book will highlight some of the challenges associated with reverse engineering of cellulose fibre bundles using biological, chemical and mechanical means.

One of the key objectives of isolating micro-and nano-cellulosic fibres by biological, chemical and mechanical means from plant and wood-based fibres is to explore their potential to give a quantum leap in performance of composite materials manufactured from them.

The Contents and Structure

First few chapters of this book focus on the isolation and characterization techniques of wood- and plant-based micro-and nano-fibres. The first chapter by Bondeson et al. presents two possible ways to prepare nanosized cellulose whiskers from microcrystalline cellulose. It demonstrates that the whiskers can form a stable water suspension by changing the surface chemistry of the crystals and that the length to diameter ratio of these whiskers is relatively high. This chapter also discusses the challenges associated with providing an environmentally friendly and economic option of making cellulose whiskers.

The second chapter by Grey and Roman documents the structure of cellulose nanocrystals from an aqueous suspension. The crystal structure is identified and the orientational order of the liquid crystalline phase is discussed. They also discuss the potential to combine these nanocrystals with polymeric resins.

The third chapter by Zimmermann et al. discusses the structure of cellulose fibril as aggregates embedded in a lignin matrix in the wood cell wall and highlights the outstanding specific tensile strength of wood. This chapter also gives some examples of producing homogeneous, translucent fibril films and polymer composites with hydroxypropyl cellulose. They demonstrate that the addition of fibrils lead to an up to three times higher modulus of elasticity and an up to five times higher tensile strength of the polymers.

Since the use cellulose whiskers, cellulose microfibrils, and cellulose fibres of micro- and nano-scale diameters as reinforcing agents in composites is increasing rapidly, this book also reviews the microstructure of cellulose in great detail. First, the structure of a cellulose microfibril in wood cell wall has been elucidated through a structural analysis of wood cell wall. Subsequently, the need to define microfibres as "fibres" for structural applications has been demonstrated.

Although the first part of this book concentrates on the isolation process of nano sized cellulose a significant effort has been made to provide an in-depth description of the state-of-art characterization techniques for nanofibre and nanocrystals. The fourth chapter by Tanem et al. gives a powerful insight into micro fibres of cellulose, cellulose whiskers and their nanocomposites with poly(lactic acid) using atomic force microscopy and electron microscopy. Transmission electron microscopy, including a bright field TEM and STEM combines to examine cellulose whiskers and microfibres. This chapter also explains the advantages of various microscopic techniques to identify and differentiate techniques for determining surface and bulk morphology of nano-scale cellulose and their nanocomposites. Limitations of these techniques are also discussed. Detailed identification of cellulose nano-structure is vital in achieving good processing properties leading to enhanced performance of the nanocomposites.

The fifth chapter by Eichhorn introduces a concept of using Raman spectroscopy for the study of cellulose nanocomposites. He has identified a shift in Raman bands to explain the molecular deformation in cellulosic nano-materials and has also identified Raman shift as a tool for monitoring the dispersion of the nanostructured phase and the level of stress-transfer.

Surface modification of cellulosic micro-, nano-fibres is identified as an essential step to resolve the challenges associated with their dispersion during their processing stages in solid state. The chapter by Renneckar et al. reviews cellulose surface modification from the standpoint of the importance of interfacial modification for the development of strong cellulose reinforcement (nanowhiskers, wood fiber, or continuous fiber). This chapter provides diverse alternative methods to modify cellulose surfaces such as esterification of pulp fibers in a nonswelling solvent; reactive processing via steam-explosion of wood; and the self assembly technique of irreversibly adsorbing amphiphilic polymers or polyelectrolytes onto cellulose surfaces.

Several chapters in this book are devoted to processing of bionanofibres in liquid and solid phases to produce bionanocomposite cast films, extruded profiles and pellets. This book also provides the state-of-art development in the surface modification of cellulose nanocrystals, effect of interface modification and the reinforcing properties of cellulose in micro- and nanofibres and, in nanocrystals forms. The chapter written by Roman and Winter discusses the effect of cellulose nanocrystals surface chemistry on the fibre-matrix compatibility.

The contribution from Mathew et al. shows novel ways to prepare cellulose-based bionanocomposites of poly (lactic acid). The chapter illustrates the important role of dispersion of cellulosic materials in extrusion process. This chapter also provides an excellent overview of the structure-property correlation of composites made with two different cellulosic substrates such as cellulose microfibrils and cellulose nanowhiskers.

Film casting technique has been widely used to demonstrate the reinforcing effect of cellulose nanocrystals in a polymeric matrix. Petersson and Oksman elegantly describe how the addition of microcrystalline cellulose will affect the transparency, mechanical, thermal and barrier properties of biopolymers. They also provide information on the production and structure of these nanocomposites. Two biopolymer matrices are used in this study, poly (lactic acid) (PLA) and cellulose acetate butyrate (CAB). Barrier properties of nanocomposites made from microcrystalline cellulose and those biopolymers are also discussed.

The chapter eleven by Nakagaito and Yano describes the production of nanocomposites exploiting the unusually good mechanical properties of cellulose microfibrils, from high-strength to optically transparent composites based on bacterial cellulose and microfibrillated cellulose from plant fibers. This chapter also gives a comparative study of the mechanical and thermal properties of cellulose nanocomposites with other substrates such as metallic and nonmetallic alloys.

In the next chapter Chakraborty et al. elucidates the pulp cell wall morphology. They propose that microfibres should have a minimum aspect ratio for adequate stress transfer from the matrix to the fibre. Consequently, microfibres are defined as cellulose strands 0.1-1 μm in diameter, with a corresponding minimum length. The reinforcing potential of cellulose microfibres is demonstrated by using them in composites with a polyvinyl alcohol (PVA) matrix.

Plant stem fibers primarily contain bundles of cellulose nanofibers with a diameter ranging between 10 to 70 nm and lengths of thousands of nanometers. In chapter thirteen Wang and Sam describe a process to extract nanofibres from soybean stock by chemo-mechanical treatments. The composition, dispersion and morphological properties of the nanofiber are investigated and their properties compared with those of hemp nanofibers. This chapter also compares the properties of nanocomposites based on thermodynamically different polymer matrices, namely polyvinyl alcohol and polyethylene. Several examples of mixing processes are introduced in this chapter which includes liquid phase film casting and solid phase melt blending. Effect of polymeric dispersant is also discussed in relation to the nanofibre dispersion and mechanical strength.

The chapter fourteen by Noorani et al exemplifies a potential use of cellulose nanocomposite in microchannel devices. Membranes made from polysulfone could be potentially nanocomposites. In this chapter the authors discuss methods to incorporate cellulose nanocrystals into polysulfone matrix using a solvent exchange process. The chapter also discusses methods to characterize such composites and their interfacial interaction by observing thermogravimetric changes.

The final chapter of this book provides an excellent review of the bacterial nanocellulose based composites. Millon et al discuss their usage in biomedical

8

applications. It is demonstrated that the bacterial cellulose fiber has a high degree of crystallinity and they are very strong. When used in combination with other biocompatible materials, produced nanocomposites are suitable for use in medical applications. In particular, nanocomposites consisting of bacterial cellulose and poly (vinyl alcohol) are investigated. A broad range of mechanical properties of composites is discussed. Materials with properties close to that of cardiovascular tissues are also reviewed. These materials are potential candidates as synthetic replacements of natural cardiovascular tissues.

Conclusions

Nanocomposites based on cellulose are a relatively new research field and until now there are no books available on this topic. There is a growing interest on biopolymer based nanocomposites in developed and developing world (Europe, Asian, North America and South America etc.) and, especially if the nanocomposites are based totally on renewable raw materials. We can see several ongoing research projects on this topic and there is an emerging research trend for developing micro-and nano-fibre reinforced biopolymers. Therefore, we gathered people who are working in this research topic to present their work at the Annual American Chemical Society meeting in San Diego on May 2005. This book is the result of that symposium.

The book deals with new nanostructured composites, where both the reinforcement and the matrix are bio based. Cellulose combined with natural polymers will lead to the development of a new class of biodegradable and environmental friendly bionanocomposites. This new family of nanocomposites shows remarkable improvement of material properties when compared with the matrix polymers or conventional micro- and macro-composite materials. Such improvements in properties typically include a higher modulus and strength, improved barrier properties and increased heat distortion temperature. This new class of renewable nanocomposites is expected to capture new market in transportation, medical and packaging applications.

This book will include the raw materials which can be used for making these composites, separation/isolation technologies of cellulose based reinforcements from different natural resources and, it will also include a brief overview of the recent advancements in the surface chemistry of nanocellulose. Characterization methods such as atomic force microscope (AFM), transmission electron microscope (TEM) and scanning electron microscope (SEM), Raman Spectroscopy, for the nano scale cellulose reinforcements and for the bio-nanocomposites will be discussed. Further, different processing methods for nanocomposites and their mechanical, thermal properties, barrier properties will be included.

Materials Characterization

Chapter 2

Strategies for Preparation of Cellulose Whiskers from Microcrystalline Cellulose as Reinforcement in Nanocomposites

Daniel Bondeson, Ingvild Kvien, and Kristiina Oksman

Department of Engineering Design and Materials, Norwegian University of Science and Technology, Rich Birkelands vei 2b, N7491 Trondheim, Norway

In this chapter we present two possible ways to prepare nanosized cellulose whiskers from microcrystalline cellulose (MCC) derived from Norwegian spruce (Picea abies). The first procedure was to prepare a stable water suspension of cellulose whiskers after treatment with sulfuric acid (H_2SO_4). In the second procedure, the whiskers were dispersed in an organic medium; dimethylacetamide/lithium chloride (DMAc/LiCl). It was possible to produce a stable colloidal suspension of cellulose whiskers with H_2SO_4. The whisker size was measured to be between 200-400 nm in length and less than 10 nm in width. The dispersion with DMAc was carried out with and without LiCl. Atomic force microscopy analysis of the sample without LiCl confirmed the presence of both whiskers and MCC particles. Characterization of the suspension containing LiCl was restricted due to difficulties in solvent removal.

Introduction

The interest in producing composite materials with nanosized reinforcement, i.e. nanocomposites, has grown tremendously in recent years. The enthusiasm is due to the extraordinary properties this kind of materials exhibit because of the nanometric size effect of the reinforcement. The challenge has been, especially for a continuous and large scale production, to get the reinforcement well dispersed and without agglomerates in a continuous matrix. Most efforts have been to produce nanocomposites with inorganic reinforcements, but organic material has also been used, including cellulose. Among the advantages of using cellulose as a renewable reinforcement is its abundance, together with easier recycling by combustion in comparison with inorganic filled systems. There are also limitations on the use of unmodified cellulose crystals due to their incompatibility with a typically more hydrophobic thermoplastic matrix and difficulties in achieving acceptable dispersion levels (1). Preparation of nanocomposites from a stable aqueous suspension is limited to either hydrosoluble polymers or an aqueous suspension of polymer, i.e. a latex to achieve a good level of dispersion (2). Several methods have been explored to achieve a dispersion of cellulose crystals, or cellulose whiskers referring to its needle like structure, in low-polar solvents to widen the possible matrixes for nanocomposite processing. The use of a surfactant (3), grafting of poly(ethylene glycol) onto whiskers (4), partially silylation of whiskers (5, 6) all led to stable suspensions in various low-polarity solvents. Azizi Samir et al (2) redispersed cellulose whiskers in an organic solvent without addition of a surfactant or any chemical modification.

Another challenge is the tedious processing steps by means of purification, bleaching, fibrillation, hydrolysis and the low yield of the final dispersion of cellulose whiskers. There are different techniques to accomplish this isolation of cellulose whiskers. Acid hydrolysis of cellulose is a well know process to remove amorphous regions and several studies have been reported where cellulose crystallites/whiskers were identified and separated from various sources. Nickerson and Habrle (7) talked about cellulose crystallites from cellulosic materials by hydrochloric and sulfuric acid hydrolysis in 1947, and in 1952 Ranby (8) reported the preparation of cellulose whiskers from microfibrils by acid hydrolysis. Marchessault with co workers studied the hydrolysis of chitin, native, mercerized, and bacterial cellulose using acid hydrolysis and reported birefringence (9,10). At CERMAV-CNRS in France, cellulose whiskers have been separated from various sources like wheat straws, tunicin etc, and have been used as reinforcements in polymer matrices (1, 11-14). Incorporation of these nanosized elements into a polymeric matrix usually resulted in outstanding properties, with respect to their conventional microcomposite counterpart. Recently, microcrystalline cellulose (MCC) has attracted attention as a possible starting material for the preparation of cellulose based nanocomposites (15). MCC is a commercially available material which is mainly used as a binder in tablets and capsules. It is prepared from native cellulose by acid hydrolysis, back-neutralisation with alkali and spray-drying (16). Strong hydrogen bonding between the individual cellulose crystals produced promotes re-aggregation during drying procedures (16). Thus, the MCC produced consists of aggregated bundles of crystallites with different particle sizes. To utilize cellulose crystals as reinforcement, the hydrogen bonds between the crystals must be broken and the cellulose crystals must be well dispersed in the matrix.

In this study we present two possible ways to prepare nanozised cellulose whiskers from microcrystalline cellulose (MCC) derived from Norway spruce (Picea abies). The aim of the first preparation technique was to prepare a stable colloidal water suspension of cellulose whiskers after treatment with sulphuric acid. In the second route the aim was to disperse the whiskers in an organic medium, DMAc/LiCl, to make the suspension compatible with low polarity polymers. This method is expected to be a new route to isolate whiskers from commercially available wood sources and obtain a stable dispersion in an organic medium, in a single step.

Sulphuric Acid

Treating the cellulose with acid, the cellulose undergoes acidic hydrolysis. It is preferably the amorphous parts of the cellulose that undergoes acidic hydrolysis rather than the crystalline (17). Hydrolysis of cellulose is greatly influenced by the acid concentration and concentrated sulfuric acid can smoothly hydrolyze the crystalline cellulose (17). Treating cellulose with sulfuric acid involves an esterification of hydroxyl groups by sulfate ions, see Figure 1 (18).

Introduction of sulfate groups along the surface of the crystallites will result in a negative charge of the surface as the pH increase. This anionic stabilization via the attraction/repulsion forces of electrical double layers at the crystallites is probably the reason for the stability of the colloidal suspensions of crystallites (10). Above critical concentrations, the suspension form a chiral nematic phase (19). It is possible to disrupt this chiral nematic phase by shear and the rods will align parallel to flow direction, exhibiting nematic ordering (20).

The cellulosic material is either directly immersed in sulfuric acid with known concentration (10,21) or immersed in water, which is kept in ice, where at sulfuric acid is added slowly to reach final concentration (12). The ice bath and slow addition of sulfuric acid is utilized to prevent elevated temperatures and to hinder hydrolysis of cellulose in the suspension during addition of acid. After this step, the suspension is heated while stirring, and it is in principal here the hydrolysis of cellulose take place. As mentioned earlier, the sulfuric acid concentration greatly influences the cellulose hydrolysis (17). Mukherjee et al.

13

Figure 1. Esterification of hydroxyl groups by sulfate ions from sulfuric acid treatment of cellulose.

(21) observed that ramie and cotton hydrolysis was effective in a sulfuric acid concentration between 9.69 and 9.94 mol/L at 20 °C in leading to a colloidal suspension. The effect of a lower concentration of acid, 9.18 mol/L, was negligible in 24 h, and relatively slight after 72 h. A concentration of 10.04 mol/L or more lead ultimately to complete solution in the acid, and if the treatment was stopped before this stage was reached, the cellulose was found to be partially transformed to Cellulose II. Along with acid concentration, time and temperature of hydrolysis are also important parameters (22). It has been reported that an increased hydrolysis time increased the surface charge and amount of sulfate groups (22).

After the hydrolysis step, the excess sulfuric acid in the suspension has to be removed and this is done by either centrifugation (10,12,19,21,23), filtration (7), or by solely dialysis (12). In centrifugation, the sediment is kept and the supernatant is removed and replaced by distilled/deionized water. The sediment is then mixed with the new distilled/deionized water and centrifuged. This procedure is continued until the supernatant is becoming turbid, often at pH between 1 and 3, and as mentioned before, this is most likely due to repulsive forces between the crystallites which come into play as the pH increases (10). The suspension is then dialyzed against distilled/deionized water to remove the last residue of the sulfuric acid, and this usually takes a couple of days. To further disperse the preparations, some minutes of ultrasonic treatment can be used (12,19,22,23). The ultrasonic treatment can be carried out in an ice bath in order to avoid overheating which might cause desulfation of the sulfate groups on the surface of the crystallites (22). Helbert et al. (12) concentrated the cellulose suspension by dialysis against high molecular weight polyethylene glycol (PEG 35 000).

Dimethyl Acetamide/Lithium Chloride

N,N-dimethyl acetamide (DMAc) containing lithium chloride (LiCl) is a well-known and favorable solvent system for cellulose (24). It was originally developed in 1976 to dissolve chitin (25). The optimum concentration of LiCl in the solvent mixture is reported to be between 5 and 9 wt% (26). Different mechanisms for the interaction between cellulose and DMAc/LiCl are suggested (27). Turbak (28) suggested that LiCl is forming a complex with dimethylacetamide, releasing Cl⁻, which acts as a base toward the hydrogen on the cellulose hydroxyl group, thus Cl⁻, plays a major role in the dissolution by breaking up the inter- and intrahydrogen bonds. It is found that the use of other salts such as LiBr, $LiNO_3$ etc does not work (28). Water has to be excluded from the solvent system since both LiCl and DMAc are very hygroscopic (27). Water hinders complexation with cellulose and promotes the formation of polymer aggregates (29). Several authors report the risk of degradation of the cellulose upon treatment with DMAc/LiCl, thus the processing conditions have to be carefully considered. According to Potthast et al *(30)* the treatment of cellulose with DMAc/LiCl might cause severe degradation of cellulose dependent of the heating time and temperature. They found that the degradation occurred via endwise peeling reactions due to N,N-dimethylacetoacetamide, which is a condensation product of DMAc formed during heating. N,N-dimethylketenium ions are formed at temperatures above 80°C and causes random cleavage of cellulose molecules. Further, they found that the degradation of pulp is strongly accelerated in the presence of LiCl. Additionally they reported a yellowing of the mixture caused by chromophores formed in LiCl-catalyzed condensation reactions from DMAc. The discoloration of the pulp also originated from furan-type structures which are formed upon heating in DMAc or DMAc/LiCl.

Apart from being an effective solvent for cellulose, DMAc/LiCl is an interesting swelling agent for cellulose. Berthold et al *(31)* reported that pure DMAc or DMAc containing only small amounts of LiCl (0,5% w/v) was sufficient for swelling of unbleached sulphate fibres which facilitated fibrillation of the fibres. It is expected that similar conditions is able to penetrate in between the cellulose whiskers in microcrystalline cellulose, which will lead to isolation of the cellulose whiskers by breaking the hydrogen bonds between the crystals.

Experimental

Materials

Microcrystalline cellulose from Norwegian spruce (Picea abies) supplied by Borregaard ChemCell, Sarpsborg, Norway, with a particle size between 10 to 15 μm, was used as starting material for preparation of cellulose whiskers.

Sulphuric acid of analytical purity was purchased from Merck (Darmstadt, Germany). DMAc of analytical purity was purchased from LAB Scan (Dublin, Ireland). Extra pure LiCl, was purchased from Merck (Darmstadt, Germany).

Methods

Microcrystalline cellulose, 10 g/100 mL, was hydrolyzed in 9mol/L sulfuric acid at 44°C in 130 mm. The excess of sulfuric acid was removed by repeated cycles of centrifugation (10 min at 12 000 rpm, Sorvall RC-5B), i.e. the supernatant was removed from the sediment and replaced by new deionized water and mixed. The centrifugation step was continued until the supernatant became turbid. The last washing step was carried out by dialysis against deionized water until the washing water was constant in pH, i.e. about neutrality.

The swelling and separation of MCC with DMAc was carried out with and without LiCl. Two suspensions of 1 wt% MCC in DMAc containing 0 and 1 wt% LiCl were prepared. The suspensions were heated and mechanically stirred for 5 days. The heating temperature was set to 60°C and 80°C for the suspensions containing 1 and 0 wt% LiCl, respectively. The suspensions were subsequently treated in an ultrasonic bath for 60 minutes. After the treatment unseparated particles were removed by centrifugation at 1000 rpm for 5 minutes. The final suspensions were stored in a refrigerator.

Characterization

Yield was calculated as % (of initial weight) of MCC after hydrolysis. For calculation of the yield after DMAc(/LiCl) treatment, a small amount of the suspensions was vacuum-dried for ~15 hours and oven-dried at 100°C for ~15 hours to remove residual DMAc(/LiCl). The dried cellulose whiskers were then kept in a desiccator before weighing.

Flow birefringence in the suspensions was investigated by using two crossed polarization filters. The samples were diluted before examination.

Optical light microscope (OM) observations were performed using a Leica DMLB. OM was used in order to detect bigger particles and to get an overview in the microscopic level. The magnifications used were × 100, ×500 and × 1000. The sulphuric acid treated sample was diluted to a concentration of 0.1 g/100 mL before examination.

Atomic force microscopy (AFM) observation was performed using a NanoScope IIIa, Multimode[TM] SPM from Veeco. Calibration was performed by scanning a calibration grid with precisely known dimensions. All scans were performed in air with commercial Si Nanoprobes[TM] SPM Tips. Height- and phase images were obtained simultaneously in Tapping mode at the fundamental resonance frequency of the cantilever with a scan rate of 0.5 line/s using j-type scanner. The free oscillating amplitude was 3.0V, while the set point amplitude was chosen individually for each sample. A droplet of the suspension was placed on a freshly cleaved mica surface and allowed to dry at 80°C over-night before analysis.

Transmission electron microscope (TEM) observations were performed using a Philips CM30 operated at 100kV. A droplet of suspension was placed on a copper grid covered by a thin carbon film and allowed to dry at 80°C overnight.

Results and Discussion

The yield after acid hydrolysis of the MCC was 30 % (of initial weight). In an earlier study it was found from wide angle X-ray (WAXD) analysis that the MCC used in this study consisted not only of crystalline cellulose but also included amorphous regions (*32*). One of the reasons why such a low yield was obtained might thus be that the residue amorphous regions in the starting material disintegrated during the hydrolysis. Further, the highest priority was to obtain a suspension of nanosized whiskers and it is possible that some of the cellulose crystals degraded during the treatment. It is also likely that cellulose whiskers were removed during the washing steps. For the DMAc(/LiCl) treated samples, the yield was only 1,5wt% for the suspension without LiCl. However, the suspension containing LiCl had a yield above 300wt%. An explanation for this unrealistic high yield is that the complex formed between the cellulose and the DMAc/LiCl system was still present in the dried sample, i.e. the DMAc/LiCl was not thoroughly removed.

Clear flow birefringence was seen for the acid hydrolyzed sample (0.1 g cellulose/100 mL) in cross polarized light, see Figure 2a. This indicates a nematic liquid crystalline alignment and reveals the existence of whiskers (*20*). The DMAc(/LiCl) suspensions appeared ivory and showed no signs of sedimentation when observed after two hours. Both suspensions showed flow birefringence, as seen in Figure 2b and c, but not as pronounced as for the hydrolyzed suspension.

A comparison of the sample prepared by acid hydrolysis with untreated MCC in OM at ×1000 magnification is shown in Figure 3. No particles could be seen for the treated sample due to restricted resolution of the light microscope. This indicates that there were possibly only nanosized cellulose whiskers in the suspension.

After treatment of MCC in DMAc(/LiCl) there were still a large number of micro-sized particles in the suspensions, but a swelling and partly separation of the MCC agglomerates had clearly occurred. The suspension without LiCl appeared to contain more scarcely separated particles than the suspension containing LiCl. It thus seemed that the solvent containing LiCl was more effective in releasing cellulose whiskers from the MCC particles. In Figure 4 OM pictures of the samples after removal of the largest particles by centrifugation are shown. The suspensions still contained a few micro-sized particles. It was however expected that in addition the suspensions contained nanosized whiskers which were not detectable in the light-microscope.

From AFM investigation of the acid hydrolyzed sample (0.1 g cellulose/100 mL), it was seen that a mat of whiskers were covering the surface after water had been permitted to evaporate, see Figure 5a. The AFM image reveals needle like structure of the cellulose, i.e. whiskers. One has to be aware of that the whiskers look thicker in AFM than in reality due to the well-known broadening effect due to tip convolution (*33*) in AFM. This will lead to a general broadening of the whiskers. Therefore, to determine the length and width of the whiskers, highly diluted samples of the hydrolyzed suspension were analysed in TEM. Figure 5b shows a TEM image of the hydrolyzed sample. Measurements from TEM images gave whiskers with a length between 200 – 400 nm and a width less than 10 nm. A tendency of agglomeration could be observed from TEM. It is not clear whether this was due to drying of the suspension or if it reflected the state of the suspension.

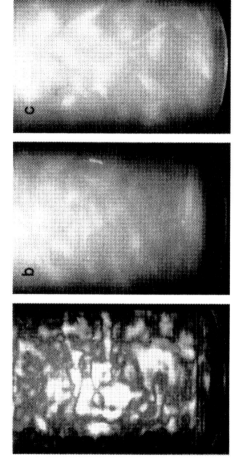

*Figure 2. Flow birefringence seen between two crossed polarizing films of the
a) hydrolyzed suspension b) DMAc suspension c) DMAc/LiCl suspension
(Container with 24 mm in diameter)*

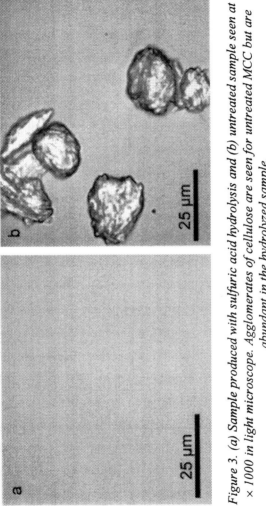

Figure 3. (a) Sample produced with sulfuric acid hydrolysis and (b) untreated sample seen at × 1000 in light microscope. Agglomerates of cellulose are seen for untreated MCC but are abundant in the hydrolyzed sample.

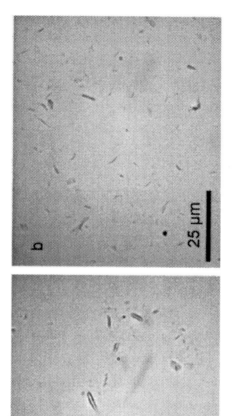

Figure 4. a) MCC treated with DMAc and b) MCC treated with DMAc/LiCl seen at x1000 in the light microscope

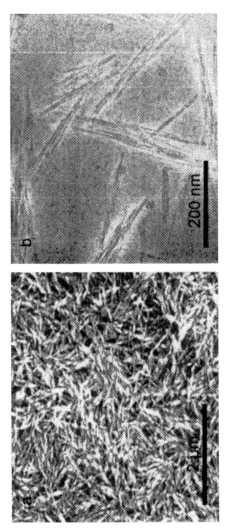

Figure 5. a) Topography AFM image of whisker suspension b) TEM image of the whisker suspension after acid hydrolysis.

Droplets of the DMAc(/LiCl) suspensions were dried onto a mica surface for observation by AFM. However, the dried sample of the suspension containing LiCl seemed to attract moisture, even after prolonged treatment in vacuum oven at 100°C. A large number of droplets appeared on the sample which made it impossible for AFM analysis. Both LiCl and DMAc are known to be very hygroscopic (27). This indicates that the DMAc/LiCl solvent was still present in the dried sample, as found for the yield calculation. The sample without LiCl, however, showed no development of water droplets and was successfully analysed by AFM. From AFM it was evident that the suspension without LiCl was not homogenous but consisted of particles at different stages of separation. This is clearly demonstrated in Figure 6a showing a tightly packed particle separating into a large number of cellulose whiskers. Due to the broadening effect it was difficult to judge whether the structures observed were individual whiskers or several whiskers, agglomerated side-by-side. Determination of the whiskers length and width was therefore uncertain. In Figure 6b it is evident that even in the sample without LiCl there was still a remainder of solvent. The underlying mica surface can be observed from a hole in the solvent film.

In addition to cellulose whiskers the picture reveals spherical particles, which are believed to be fragments of cellulose. However, these fragments were already present in the MCC before DMAc was added. These fragments are presumably resulting from the sulphuric acid treatment of cellulose in the production of MCC. It is expected that the sample with LiCl contained more cellulose whiskers compared to the sample without LiCl due to the higher yield and visible water uptake from the dried sample. This was however not possible to confirm by AFM analysis.

Conclusions

Two possible ways to prepare cellulose whiskers from commercially available microcrystalline cellulose (MCC) based on wood were explored. The aim of the first method was to prepare a stable water suspension of cellulose whiskers after treatment with sulfuric acid. In the second method the aim was to disperse the whiskers in an organic medium, DMAc/LiCl, to make it compatible with low polarity polymers.

With sulfuric acid concentration of 9 mol/L, it was possible to produce cellulose whiskers with length between 200 – 400 nm and a width less than 10 nm in approximately 2 h and with a yield of 30 % (of initial weigth). Disintegration of amorphous regions and degradation of crystalline parts during hydrolysis, and lost via the washing steps is probably the explanation for the low yield.

The swelling and separation with DMAc was carried out with and without LiCl to study the effect of LiCl in the separation process. Both suspensions appeared ivory and showed birefringence between cross polarizers, which is an

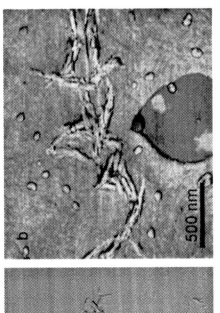
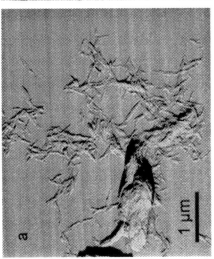

Figure 6. a) AFM phase image of the dried suspension without LiCl showing both cellulose particles and cellulose whiskers b) AFM phase image of the dried suspension without LiCl showing remainder of solvent

indication of the presence of whiskers. The solvent containing LiCl appeared to be more effective in releasing the cellulose whiskers from the MCC agglomerates. However, characterization of the suspension containing LiCl was challenging due to difficulties in removing the solvent. This was reflected in the unrealistic high yield of the sample containing LiCl. Atomic force microscopy analysis of the sample without LiCl confirmed the presence of cellulose whiskers, but also revealed unseparated MCC particles.

Acknowledgements

We thank the Research Council of Norway for financial support and Borregaard ChemCell in Sarpsborg, Norway, for providing us with MCC. We also thank Kristin Brevik Antonsen and Dr. Simon Ballance at Department of Biotechnology at Norwegian University of Science and Technology for providing us with necessary equipment used in this research.

References

1. Dufresne, A.; Cavaillé, J.Y. ACS Symp. Ser. **1999**, *723*, 39-54
2. Azizi Samir, M.A.S.; Alloin, F.; Sanchez, J-Y.; Kissi, N.E.; Dufresne, A. Macromolecules **2004**, *37*, 1386-1393
3. Heux, L.; Chauve, G.; Bonini, C. Langmuir **2000**, *16*, 8210-8212
4. Araki, J.; Wada, M.; Kuga, S. Langmuir **2001**, *17*, 21-27
5. Goussé, C.; Chanzy, H.; Excoffier, G.; Soubeyrand, L.; Fleury, E. Polymer **2002**, *43*, 2645-2651
6. Goussé, C.; Chanzy, H.; Cerrada, M.L.; Fleury, E.; Polymer **2004**, *45*, 1569- 1575
7. Nickerson, R.F.; Habrle, J.A. Ind. Eng. Chem. **1947**, *39*, 1507-1512
8. Ranby, B.G. Tappi **1952**, *35*, 53-58
9. Marchessault, R.H.; Morehead, F.F.; Walter, N.M. Nature **1959**, *184*, 632-633
10. Marchessault, R.H.; Morehead, F.F.; Koch, M.J. J. Colloid Sci. **1961**, *16*, 327-344
11. Favier, V.; Chanzy, H.; Cavaillé J.Y. Macromolecules **1995**, *28*, 6365-6367
12. Helbert, W.; Cavaillé, J.Y.; Dufresne, A. Polym. Comp. **1996**, *17*, 604-610
13. Angles, M.N.; Dufresne, A. Macromolecules **2000**, *33*, 8344-8353
14. Morin, A.; Dufresne, A. Macromolecules **2002**, *35*, 2190-2199
15. Mathew, A.P.; Oksman, K.; Sain, M. *J. Appl. Polym. Sci.* **2005**, *97*, 2014-2025
16. Levis S. R.; Deasy P.B. Int. J. Pharm. **2001**, *213*, 13-24
17. Hon, D. N.-S.; Nobuo, S. Wood and Cellulosic Chemistry, Marcel Dekker, Inc., New York **1991**, pp 997-1013

18. Yao, S. Chin. J. Chem. Ing. **1999,** *7,* 47-55
19. Revol, J.-F.; Bradford, H.; Giasson, J.; Marchessault, R.H.; Gray, D.G. Int. J. Biol. Macromol. **1992,** *14,* 170-172
20. Orts, W.J.; Godbout, L.; Marchessault, R.H.; Revol, J.-F. Macromolecules **1998,** *31,* 5717-5725
21. Mukherjee, S.M.; Woods, H.J. Biochim. Biophys. Acta **1953,** *10,* 499-511
22. Dong, X.M.; Revol, J.-F.; Gray, D.G. Cellulose **1998,** *5,* 19-32
23. Araki, J.; Wada, M.; Kuga, S.; Okano, T. Colloids Surf. A: Physiochem. Eng. Aspects **1998,** *142,* 75-82
24. Striegel, A.M. Carbohydr. Polym. **1997,** *34,* 267-274
25. Austin, P.R. Ger. Offen. U.S. Patent 4,059,457, 1977
26. Dawsey, T. R.; McCormick, C.L. J. Macromol. Sci, Rev. Macromol. Chem. Phys. **1990,** *C30,* 405-40.
27. Dupont A.L. Polymer **2003,** *44,* 7-17
28. Turbak, A.F. Tappi J. **1984,** *67,* 94-96
29. Potthast, A.; Rosenau, T.; Buchner, R.; Roeder, T.; Ebner, G.; Bruglachner, H.; Sixta, H.; Kosma, P. Cellulose **2002,** *9,* 41-53
30. Potthast A.; Rosenau T.; Sartori J.; Sixta H.; Kosma P. Polymer **2003,** *44,* 7-17
31. Berthold, F.; Sjöholm, E.; Gustafsson, K.; Gamstedt, E.K.; Salmén, L. In Sustainable Natural and Polymeric Composites – Science and Technology; Lilholt, H.; Madsen, B.; Toftegaard, H.L., Cendre, E.; Megnis, M.; Mikkelsen, L.P.; Sørensen, B.F., Ed.; Risø National Laboratory; Roskilde, DK, 2002
32. Kvien, I.; Tanem, B.S.; Oksman, K. Proceeding, 8th Int. Conference on Woodfiber-Plastic Composites, Wisconsin, USA, 2005
33. Markiewiez P.; Goh M.C. J. Vac. Sci. Technol. **1995,** *B13,* 1115

Chapter 3

Self-Assembly of Cellulose Nanocrystals: Parabolic Focal Conic Films

Derek G. Gray[1] and Maren Roman[2]

[1]Department of Chemistry, Pulp and Paper Building, McGill University, Montréal, Québec H3A 2A7, Canada
[2]Department of Wood Science and Forest Products, Virginia Polytechnic Institute and State University, Blacksburg, VA 24061

Evaporation of aqueous suspensions of cellulose nanocrystals gives solid films which retain the orientational order of the liquid crystalline phase observed in the liquid state. At a given film thickness, a parabolic focal conic texture is fixed in the solid film. A brief explanation of this structure, and a tentative rationale for its formation are presented.

Introduction

In nature, cellulose is the main structural polymer in the plant cell wall, where it exists as partially crystalline fibrous microfibrillar aggregates. Under suitable conditions, acid hydrolysis breaks down the aggregates into individual needle-shaped crystalline rods of colloidal dimensions (*1*) whose length and breadth depend on the cellulose source and the hydrolysis conditions. The width of the rods is of the order of a few nanometers, and so the resultant highly crystalline cellulose I particles are referred to as nanocrystals. It is convenient to

26

distinguish between these nanocrystals, whose length is of the order of tens or hundreds of nanometers, and the much longer elementary fibrils or microfibrils of cellulose that are the primary biosynthesized species in plant cell walls.

When suitably stabilized, aqueous suspensions of cellulose nanocrystals may form chiral nematic liquid crystalline phases (2). When such a suspension is allowed to dry down slowly on a flat surface, the chiral nematic order of the crystals is preserved in the resulting solid film. The iridescent properties of these films suggest decorative and security applications (3). The basic structure of these films depends on the self-orientation of the nanocrystals into the well-known twisted or helicoidal arrangement. However, this idealized arrangement is seldom observed for thick samples. To elucidate the arrangement of nanocrystals in these self-assembled films, we investigated films of different thickness by polarized-light and atomic force microscopy.

Experimental

Cellulose nanocrystals were prepared from dissolving-grade softwood sulfite pulp lapsheets (Temalfa 93, Tembec Inc., Temiscaming, QC, milled in a Wiley mill to pass a 20-mesh screen. The milled fibers were hydrolyzed for 45 mm at 45 °C with 8.75 mL of 64 wt% sulfuric acid per gram of cellulose. The hydrolysis was quenched by diluting 10-fold with cold water. The crystals were collected and washed once by centrifugation for 10 min at 5000 rpm and then dialyzed against ultrapure water (Millipore Milli-Q UF Plus) until the pH was neutral. Crystal aggregates were disrupted by sonication. The suspension was kept over ion exchange resin for 4 days and then filtered through Whatman 541 filter paper. The final concentration of the suspension was 0.7 wt.%. Solid films of the nanocrystals were prepared by drying down 10 mL of the suspension under ambient conditions in small polystyrene Petri dishes (50 mm dia.). The thickness of the films was measured by optical microscopy on film cross-sections.

Results and Discussion

Figure 1 shows images of films of different thickness cast from aqueous cellulose nanocrystal suspensions. On the left, the thinnest film is relatively featureless, showing only a few isolated disclination lines (4). This optical texture is characteristic of a planar texture, where the helicoidal axis is normal to the plane of the film. The thickest sample, on the right in Figure 1, shows a more complex texture resembling the polygonal focal texture often observed for smectic liquid crystals (4).

28

Figure 1. Polarizing microscope images (area 300 μm x 300 μm) of cellulose nanocrystal films, thickness (left to right), 12 μm, 24 μm, 36 μm. The parabolic focal conic texture is shown in the centre image.

In films of intermediate thickness (Figure 1, centre image), we unexpectedly observed regular arrays of parabolic focal conics (PFCs), a special case of focal conic defect structures (5). PFCs are well known in thermotropic smectic phases of small molecules and polymers, and in lyotropic lamellar phases of low-molecular-weight and polymeric surfactants and lipids. To our knowledge, this is the first occurrence of the PFC structure in a colloidal liquid crystal. In addition, our solid films offer a unique opportunity for the understanding of these defects. Here, we indicate qualitatively the nature of the film structure, speculate on why it forms, and suggest its significance.

Why parabolic?

Imagine, for the moment, that the chiral nematic structure can be viewed simply as a layered fluid structure, where the layers correspond to a parallel orientation of the nanocrystal orientation in the layer. The layers will tend to arrange themselves parallel to a neighbouring flat surface, such as a microscope slide and a cover slip.

In Figure 2, the relationship between flat and curved planes in a parabolic focal conic defect is sketched in cross-section. A parabola, defined as the set of points in a plane that are equidistant from a fixed point (the focus) and a fixed line (the directrix), is found at the intersection between a set of flat parallel and nested spherical planes. The orientation is indeterminate at the line of defects, which thus appears dark under the polarizing microscope.

The parabolic focal conic arrangement consists of confocal pairs of upward and downward facing parabolas at right angles to each other. Viewed from above along the z-direction, these appear as dark crosses, as shown in the center image of Figure 1. A full description of defects and focal conic textures in liquid crystals is given in many texts (6, 7).

Why does the parabolic focal conic texture form?

The formation of a regular parabolic focal conic liquid crystalline texture during the evaporation of water from the chiral nematic suspension of cellulose nanocrystals is surprising. We speculate that the change in volume on evaporation from the liquid surface in the Petri dish causes compressive strain within the liquid that results in layer buckling. The situation is sketched in Figure 3.

30

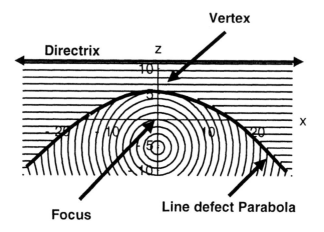

Figure 2. Sketch of the parabola-shaped intersection between a set of flat parallel and nested spherical planes.

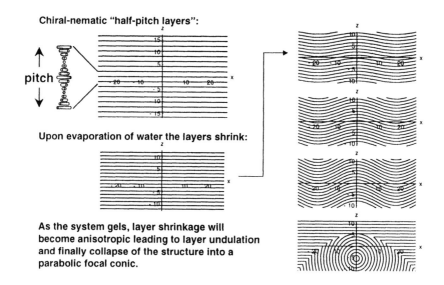

Figure 3. Sketch of possible sequence leading to formation of a parabolic focal conic texture from a planar texture(5). The numbers on the axes correspond to dimensions in μm for a typical chiral nematic cellulose film.

Most focal conic structures have been observed for smectic liquid crystals, where the fluid layered structure is evidently conducive to this type of defect formation. The "layering" of chiral nematics is less well-defined; the fingerprint texture observed for chiral nematics in general (and also in the cross-section of these cellulose films) is due to the alternating refractive index orthogonal to the helicoidal axis as the nanocrystals are oriented along and across the viewing direction. But the layers in chiral nematics are not real; there is a smooth gradation of properties normal to the helicoidal axis. So other factors may be involved in the formation of this texture from aqueous colloidal suspensions of cellulose nanocrystals.

Concluding Comments

Our PFC films demonstrate that, starting from a simple rod-like colloid, it is possible to spontaneously form regular and complex structures that can be trapped in a solid film. The dimensions of the observed defect structure in these cellulose nanocrystal films have been fitted to a calculated model of parabolic focal conic structure (*8*).

The film structure is not purely of academic interest; the iridescence observed from these films depends both on the pitch length of the chiral nematic structure (which governs the basic reflection colour) and also on the texture. An ideal planar texture reflects a single sharp wavelength of light at a given angle. The parabolic focal conic texture reflects over a range of angles and wavelengths, giving a more sparkling reflection.

The solid film is essentially homogeneous, being pure Cellulose I. The structure is due to the variation in nanocrystal orientation in different regions of the film. However, the film material may be combined or embedded in a matrix such as an epoxy resin (*3*) to enhance optical and material properties.

Acknowledgements

We thank the Natural Sciences and Engineering Research Council of Canada for financial support and the Center for Self-Assembled Chemical Structures. FQRNT (Québec), for infrastructure support.

References

1. Mukherjee, S. M.; Woods, H. J. *Biochim. Biophys. Acta* **1953**, *10*, 499–511.
2. Revol, J. F.; Bradford, H.; Giasson, J.; Marchessault, R. H.; Gray, D. G. *Int. J. Biol. Macromol.* **1992**, *14*, 170–172.

3. Revol, J.-F.; Godbout, J. D. L.; Gray, D. G. U.S. Patent 5,629,055, 1997.
4. Collings, P.; Hird, M. *An Introduction to Liquid Crystals: Chemistry and Physics;* The Liquid Crystals Book Series; Taylor & Francis: London, UK, 1997.
5. Rosenblatt, C. S.; Pindak, R.; Clark, N. A.; Meyer, R. B. *J. Phys. (Paris)* **1977**, *38,* 1105–1115.
6. Chandrasekhar, S. *Liquid Crystals;* 2nd ed.; Cambridge University Press: Cambridge, UK, 1992.
7. Bouligand, Y. In *Physical Properties of Liquid Crystals;* Demus, D.; Goodby, J.; Gray, G. W.; Spiess, H. W.; Vill, V., Eds.; Wiley-VCH: New York, NY, 1999; pp 304–351.
8. Roman, M.; Gray, D. G. *Langmuir* **2005**, *21*, 5555–5561.

Chapter 4

Cellulose Fibrils: Isolation, Characterization, and Capability for Technical Applications

Tanja Zimmermann[1], Evelyn Pöhler[1], Thomas Geiger[2],
Jürg Schleuniger[2], Patrick Schwaller[3], and Klaus Richter[1]

[1]Wood Laboratory, [2]Laboratory for Functional Polymers, and [3]Laboratory for Materials Technology, Empa, Swiss Federal Laboratories for Materials Testing and Research, Duebendorf and Thun, Switzerland

Cellulose fibril aggregates embedded in a lignin matrix in the cell wall are a predominant reason for the outstanding specific tensile strength of wood. In order to convert these mechanical properties to practical use for polymer composites, the fibrils can be isolated out of sulphite pulp. The obtained fibrils have diameters below 100 nanometer and lengths in the micrometer range. Homogeneous, translucent fibril films and polymer composites with hydroxypropyl cellulose can be prepared. For mechanical characterization tensile tests and nanoindentation experiments were carried out. The addition of fibrils led to an up to three times higher modulus of elasticity and an up to five times higher tensile strength of the polymers. Network formation was identified by Transmission Electron and Atomic Force Microscopy in films with a filling ratio of at least 10 %. The perspectives of producing new bio-based nanomaterials are promising.

33

Introduction

The exploitation of lignocellulosic fibres as load bearing constituents in composite materials has been investigated by a number of researchers (*1, 2*). Fibres have diameters in the micrometer and lengths in the millimeter range. They combine good mechanical properties with low density, biodegradability and renewability.

Also cellulose fibrils with diameters in the nanometer and lengths in the micrometer range become of more and more interest as reinforcing components in different polymer matrices. They have approximately a two times higher modulus of elasticity, a higher aspect ratio and a better compatibility with matrix materials compared to fibres. Earlier studies report on cellulose whiskers or fibrils obtained from organic materials like sugar beet pulp (*3-5*), potato tuber cells (*6, 7*), wheat straw (*8, 9*), tunicin (*10-12*), or crab shell chitin (*13*). These whiskers or fibrils were used as reinforcement components in synthetic polymers and biopolymers for the production of films and lacquers. A review on such cellulose based nanocomposites is given by Berglund (*14*). The possible fields of application of cellulose fibrils with high aspect ratios separated from wood fibres were hardly investigated. Thereby, wood as basis material for the extraction of cellulose fibrils offers advantages like high strength, low costs and constant availability.

Recent studies of the Wood Laboratory at the Swiss Federal Laboratories for Materials Testing and Research concentrated on the isolation of cellulose fibrils from sulphite wood pulp, their reinforcing potential as well as the characterization of fibrils and polymer composite films (*15, 16*). Some of the obtained results will be discussed in this article.

Materials and Methods

Separation of Cellulose Fibrils

As basis material sulphite pulp (mainly composed of softwood tracheids) from the Company Borregaard was used. For the separation 5 g of wet sulphite pulp with a dry content of 30 % was dispersed in 300 ml of deionised water. The suspension was treated with an ultra-turrax (FA IKA; 24000 rpm, 8 h) at 5-10° C to separate the fibril bundles from the wooden cell wall. A dispersing and homogenisation of the cellulose fibrils was achieved by application of a microfluidizer M-110 y (FA Microfluidics; 1000 bar, 60 min).

Production of Polymer Composite Films

Low viscosity hydroxypropyl cellulose (HPC, Mw= 80000 g/mol, FA Aqualon) was solved in the aqueous fibril suspensions with different solid contents (1, 5, 10 and 20 wt%) at a temperature of 60 °C. The obtained suspensions were casted in silicone forms and dried at standardized climate (23 °C/50 % r.H.) for seven days (solution casting method). Additionally, films out of pure cellulose fibril suspensions and pure HPC were also prepared.

FE – SEM Characterization of Fibrils

For the preparation of samples for Field Emission Scanning Electron Microscopy (FE-SEM, DB235S-FEG), glimmer plates were fixed with a conducting carbon on a specimen holder. A drop of a diluted fibril/water suspension (1:20) was put on the glimmer plate. The samples were air-dried and the remaining fibrils were sputtered with a platinum layer of about 5 nm (BAL-TEC MED 020 coating system). The images were taken with an accelerating voltage of 18 kV.

TEM Characterization of Fibrils and Composite Films

For investigations of isolated fibrils uncoated 300 mesh copper grids were drawn through the fibril suspension (see above). The sticked fibrils were stained with 1 wt% uranyl acetate.

For investigation of a HPC- composite film with a fibril content of 20 wt%, a small extract (2 x 2 mm^2) was embedded in Polymethylmethacrylate (PMMA) resin. Ultra thin sections (approx. 60 nm thick) of the transverse film surfaces were sectioned using an ultramicrotome fitted with a diamond knife. The sections were mounted on Formvar coated grids and examined with a Philips CM200 transmission electron microscope at an accelerating voltage of 120 kV.

AFM Characterization of Composite Films

Small pieces (5 x 5 mm^2) of the composite films of all fibril portions were glued (superglue, Turbo Klebstofftechnik GmbH, Bazenheid, CH) onto aluminium sample holders and examined with a NanoScope IVa Dimension^{TM} 3100 AFM using Tapping Mode^{TM}. The samples were scanned using a silicon tip with a radius of approx. 10 nm. The length of the cantilever was 124 µm, the spring constant was 37 N/m, and the resonant frequency was 300 kHz. The samples were investigated at ambient temperature and humidity. Images were taken in height mode, where the deflection of the cantilever is directly used to measure the z position, and in phase mode, where the phase shift of the cantilever is used to determine differences in material constitution.

Tensile Tests

Tensile tests were performed on the casted films according to EN ISO 527-4 (*17*) using a Zwick Z010 universal testing machine with a 200 N load cell. The tests were carried out at 20 °C/65 % r.H. and with a loading speed of 50 mm/min. The strain was determined by the longitudinal motion of the testing machine.

Nanoindentation

In a typical nanoindentation experiment a diamond indentation body is pressed into the specimen. The applied normal load P and the displacement into the surface h are continuously measured during loading and unloading (*18*). A schematic example for such a load-displacement curve is shown in Figure 1.

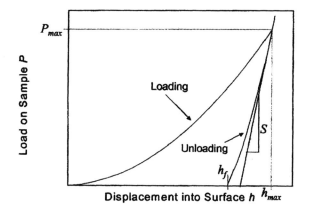

Figure1. Typical example of a load-displacement curve; h_f=residual depth of the indent after complete unloading, S= unloading stiffness dP/dh. (Reproduced with permission from reference (16). Copyright 2005.)

The Modulus of Elasticity (or indentation modulus) E_{Indent} can be calculated from the unloading part of the curve as the unloading is a purely elastic recovery process. The full procedure for the determination of E_{Indent} is described by (*19*).

The nanoindentation experiments were performed at ambient temperature using a MTS Nanoindenter XP (MTS, Oak Ridge TN, USA). Small pieces (10 x

10 mm^2) of each composite (containing 1, 5, 10 and 20 wt% of cellulose fibrils) as well as pure HPC and pure fibril films were fixed with superglue (Turbo Klebstofftechnik GmbH, Bazenheid, CH) on aluminium specimen holders. The load-displacement curves have been measured with a peak load of 12 mN, the loading rate was 0.4 mN/sec, the unloading rate 0.8 mN/sec. The peak load was kept constant for a period of 15 sec to account for possible creep effects of the sample material. After unloading 90% of the peak load, a second hold segment (i.e. keeping a constant load) of 50 sec was performed to measure the thermal drift. The latter was used to correct the load-displacement data. For each sample 50 (100 for the HPC samples with a fibril portion of 20 %) single load-displacements were measured and the average modulus values as well as the standard deviations were calculated. The large number of tests was necessary because the samples are laterally not homogeneous. Independent T-tests were carried out to demonstrate differences in means. A sample Poisson number of 0.26 was used for the calculation of the Modulus of Elasticity.

Results and Discussion

FE-SEM and TEM Characterization of Cellulose Fibrils

Mechanical disintegration of pulp fibres as described above leads to fibril structures with diameters between 20 and 100 nm. The estimated lengths are in the range of several tens of micrometers. Taniguchi and Okamura (20) also used wood pulp and a super-grinding method to obtain fibrils with diameters in the range of 20 to 90 nm. The method consists of a simple mechanical treatment which is designed to expose shearing-stress to the longitudinal fibre axis of the fibre samples. The same principle acts in our study by using a microfluidizer that separates the fibrils under high pressure and therefore also applies shearing-stress to the fibre axis. The resulting fibril dimensions are similar in both methods. Turbak et al. (21) and Herrick et al. (22) also obtained fibrilar structures with diameters in the range of 25 – 100 nm from wood pulp after a solely mechanical disintegration process.

Due to the expected high density of hydroxyl groups at the fibril surface, they strongly interact and tend to agglomerate. Thus, a network of single fibril filaments respectively superposed fibrils becomes visible on SEM and TEM micrographs (Figures 2 and 3). The importance of network formation for a high reinforcement efficiency even with small amounts of cellulose is discussed in many studies carried out with fibrils obtained from different organic fibers (not wood pulp) (8, 9, 23).

38

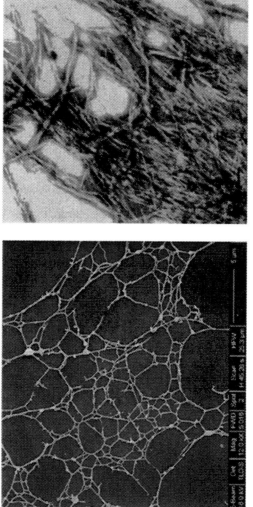

Figures 2 + 3. FE-SEM and TEM micrograph: Mechanically isolated cellulose fibrils and fibril bundles out of sulphite pulp. A fibril network with single filaments with thicknesses in the nanometer range and estimated lengths of several tens of micrometers becomes visible.

TEM and AFM Characterization of Composite Films

TEM micrographs of transverse surfaces of the HPC composite with a fibril content of 20 wt% show lighter and darker structures that can be explained by a contrast of crystalline fibril areas and the amorphous polymer matrix. The fibrils are homogeneously dispersed within the polymer matrix and form networks (Figure 4). According to the size of the fibrils, the transparency of the film also confirms the good fibril dispersion, i.e. no bigger aggregates are present. This indicates that during the drying step no fibril agglomeration occurs even when the stabilization by the fibril suspension is broken down.

So far morphological characterizations have been carried out by TEM to measure the dimensions of cellulose whiskers or fibrils (*3, 7, 9, 10, 15, 24-28*). Scanning electron microscopy (SEM) was performed for investigation of the surface morphology of fractured nanocomposite films (*7-10, 12, 15, 21, 29*). The objective of these studies was to analyse the filler/fibril distribution within the polymer matrices. From the results obtained it has been assumed that transversal sections of cellulose fibrils or whiskers appear as white dots, and the authors concluded that the filler must be homogeneously dispersed within the matrix. Thus, the dispersion of the fibrils could be shown but not the apparent network formation within the polymer matrices as demonstrated in our study for the first time. Evidence for network formation was only indirectly derived from the mechanical and physical behaviour of the composites (*7, 24*), for example from the reduced swelling behaviour of filled polymers (*10*).

Favier et al. (*30*) used TEM for the investigation of films consisting of cellulose whiskers obtained from tunicin and latex. They found that the whiskers were distributed throughout the structure, without segregation or association. Also Ruiz et al. (*31*) characterized the microstructure of nanocomposites consisting of tunicate and epoxy resin with TEM. In accordance to our results, their micrographs showed a good dispersion of the whiskers within the epoxy matrix and equal dimensions similar to the initial whisker suspension.

Only few studies report on characterization of cellulose nanocomposites by AFM. Matsumara and Glasser (*32*) evaluated the phase dimensions of thermoplastic cellulose composites by AFM. Some investigations used AFM for the characterization of pulp fibres (*33, 34*). Even though some limitations of AFM are mentioned in these studies, the authors demonstrated the usefulness of this method.

The AFM phase images obtained in our study give information about the fibril/polymer composition on the surfaces of the casted films, but in contrast to the TEM investigations no information about the distribution of the fibrils across the thickness of the films. Lower magnifications (e.g. 25 μm scans) of hydroxypropyl cellulose films with fibril contents between 1 wt% and 20 wt% show that the fibrils are quite well dispersed within the matrix. However,

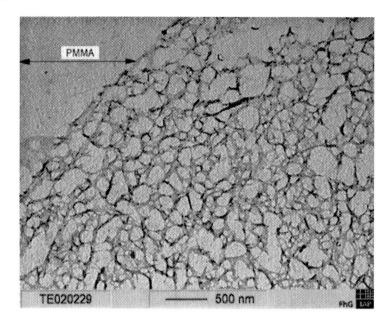

Figure 4. TEM micrograph: Transverse section of a HPC film with a fibril content of 20 wt%. The fibrils are homogeneously dispersed within the polymer and form a conspicuous network structure. PMMA: Polymethylmethacrylate. (Reproduced with permission from reference (16). Copyright 2005.)

structural irregularities and inhomogeneities with denser and more translucent areas were also detected. Higher magnifications of the composites with a fibril loading of 10 wt% and 20 wt% show network structures without indication for fibril agglomerations (Figure 5) and support the TEM findings. For the films with a fibril content of 1 wt% or 5 wt% no network formation could be detected. The fibril concentrations seem to be too small for intensive interactions between single filaments.

According to the results obtained, the direct use of the aqueous fibril suspensions for the compounding with HPC was an effective method of dispersing the fibrils within the polymer matrix. Chakraborty et al. (35) obtained similar results for microfibrils in water suspension used to make a biocomposite with polylactic acid (PLA). After a mixing procedure at 190 °C, they got a uniform dispersion of the microfibrils in the matrix.

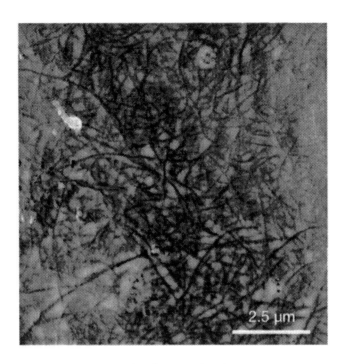

Figure 5. AFM micrograph: The fibrils are locally, homogeneously dispersed without any irregularities or agglomerations.
(Reproduced with permission from reference (16). Copyright 2005.)

Tensile Tests and Nanoindentation

The fibril/polymer suspensions could be casted to homogeneous and translucent films with thicknesses between 0.04 and 0.1 mm. Pure fibril films had thicknesses around 0.01 mm.

The results of the tensile tests are presented in Figures 6 and 7. Even though the fibrils are oriented at random in the matrix material, the Modulus of Elasticity (MOE) increases for the composites up to a factor of 3 (fibril content 20 wt %) compared to the unfilled polymer. The tensile strength even showed an up to fivefold increase. Pure films of cellulose fibrils are reaching almost the strength properties of clear wood (Figure 6).

The nanoindentation experiments led to 2-fold - 3-fold (significance level $p< 0.001$) higher MOE values for the composites when compared with those obtained in tensile tests (Figure 7). This indicates an apparent difference in the deformation behaviour between tensile and compressive loading and confirms results of other studies (36-38). Three possible factors that could lead to higher MOE values in nanoindentation experiments than calculated in tensile tests are intensively discussed in Zimmermann et al. (16):

• the differences in terms of the hydrostatic stress component which enhances the compressive modulus values in the isotropic polymer portion
• the fact that tensile and indentation tests do not probe the same material volumes and regions
• the difference in strain rate between displacement-controlled tensile tests and nanoindentation

An uncertainty of the tensile tests is related to the determination of the elongation by the longitudinal motion of the testing machine. As considerable higher moduli were obtained by nanoindentation it is reasonable that due to the shift of the tensile testing machine, the measured strain was too high for all composite films and therefore the calculated MOE values were too low. For an accurate optical strain measurement the application of a video-extensometer will be inalienable in future studies.

For the pure fibril films almost equal MOE values were measured with both methods. The reason could be related to the extremely brittle nature of these films with almost no elongation at fracture and therefore only a small influence of the strain measurements.

As it was our objective in these studies to derive MOE values to relatively compare the stiffness of the fibril reinforced composite films among themselves, the accuracy of the single numerical values was therefore negligible.

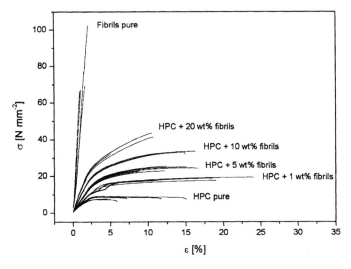

Figure 6. Stress-strain diagram of films out of hydroxypropyl cellulose and different portions of mechanically isolated cellulose fibrils. (Adapted with permission from reference (15). Copyright 2004.)

For nanoindentation experiments the maximum stiffness was found for HPC with a fibril content of 10 wt%; these films had 2 times higher indentation moduli compared to the pure polymer. The highest increase in MOE (about 35 %) was found between a fibril loading of 5 wt% and 10 wt% (p< 0.001). In comparison, films with a fibril portion of 20 wt% showed slightly lower MOE values (p< 0.05). On the other hand, the highest MOE values measured by conventional tensile testing were obtained for the films with a fibril loading of 20 wt%, in this case even 3-fold higher values compared to pure HPC. Thus, the optimal filling rate of HPC is determined between 10 wt% and 20 wt%.

This is in good agreement with results of Borges et al. (*39*) who studied the tensile properties of cross-linked (with 1,4-butyldiisocyanate) and uncross-linked composite films prepared from HPC with incorporation of microcrystalline cellulose fibres. The concentration of fibres in the composites ranged from 0 to 30 wt%. Maximum values of the MOE were observed at 10 wt% for the crosslinked and 20 wt% for the uncross-linked composites.

However, the absolute strength and MOE values differed from those measured in our study. For HPC with a fibril content of 10 % Borges et al. determined a tensile strength of 8.14 N/mm² for uncross-linked and 13.2 N/mm² for cross-linked films, e.g. The HPC films with 10 wt% fibrils of our study had an explicitly higher tensile strength of 29.7 N/mm². The different results might be an indication for a better compatibility with the matrix and a higher reinforcement potential of nanoscaled cellulose fibrils compared to the larger microcrystalline cellulose fibres.

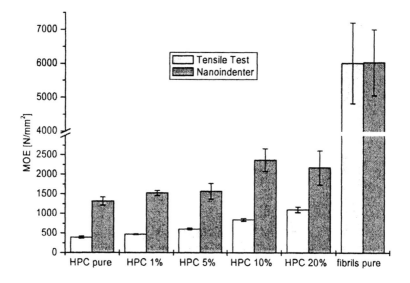

Figure 7. Comparison of the Modulus of Elasticity of the hydroxypropyl cellulose composites with different fibril portions determined in tensile tests and from nanoindentation. (Reproduced with permission from reference (16). Copyright 2005)

According to the results of our studies (*15, 16*), cellulose fibrils could be very useful for an application in water-borne coatings or adhesives to systematically improve their mechanical properties (hardness, tensile strength, MOE), their application and exploitation as well as thermal creep. In future, it is conceivable that characteristics of different matrices (e.g. fire resistance) could be directly influenced by functionalizing the hydroxyl groups of cellulose fibrils.

The microscopic characterizations of the HPC films with different fibril contents provide valuable information for the interpretation of the results obtained from nanoindentation and, respectively in tensile testing. A significant increase in the mechanical properties was obtained for the films with fibril contents of 10 wt% compared to the films with lower fibril portions. As a network formation could only be observed for composites with a fibril portion of at least 10 wt% it seems clear that the existence of fibril networks is of great importance for the mechanical properties of nanocomposites. The findings support earlier discussions about the importance of percolating cellulose fibril or whisker networks for a high reinforcement efficiency (8, 9, 15, 23).

Conclusions

From the results it is concluded that cellulose fibrils with diameters below 100 nm can be isolated out of sulphite pulp by solely mechanical processes.

The high Modulus of Elasticity (MOE) as well as the good aspect ratio of cellulose fibrils are ideal requirements for a reinforcing application in polymers. Compared with tensile testing, nanoindentation is a suitable method for mechanical characterization of wood based nanocomposites. With small sized samples the handling is much easier than for conventional testing facilities and the results seem to be reliable. The apparent differences in the deformation behaviour between tensile (conventional testing) and compressive (nanoindentation) loading require further investigation.

TEM und AFM are very useful for the morphological characterization of cellulose/HPC-composites. At higher magnification, both methods indicated a homogeneous distribution of cellulose fibrils. Network formation within the polymer for fibril contents of at least 10 wt% could be shown for the first time. Nevertheless, on a larger scale some irregularities were apparent. These inhomogeneities may explain the partly high standard deviations of the mechanical testing results.

The combination of the methods applied in this study allowed to explain the mechanical behaviour of the HPC nanocomposites and to determine the optimal filling rate for cellulose fibrils.

Acknowledgements

We thank K. Weiss (Empa, Wood Laboratory) for his support on the evaluation of the data. We are also grateful to P. Gasser (Empa, Concrete /Construction Chemistry Laboratory) and M. Pinnow (Dept. of Structural Characterization, Fraunhofer Institute Golm, D) for their electron microscopical help. Last but not least we appreciated very much the hospitality of the University of Basel (P. Reimann, V. Thommen) that make our AFM investigations possible.

46

References

1. Bledzki, A. K.; Gassan, J. *Progress in Polymer Science* 1999, *24*, 221-274.
2. Eichhorn, S. J.; Baillie, C. A.; Zafeiropoulos, N.; Mwaikambo, L. Y.; Ansell, M. P.; Dufresne, A.; Entwistle, K. M.; Herrera-Franco, P. J.; Escamilla, G. C.; Groom, L.; Hughes, M.; Hill, C.; Rials, T. G.; Wild, P. M. *J. Mater. Sci.* 2001, *36*, 2107-2131.
3. Dinand, E.; Chanzy, H.; Vignon, M. R. *Food Hydrocolloids* 1999, *13*, 275-283.
4. Heux, L.; Dinand, E.; Vignon, M. R. *Carbohydrate Polymers* 1999, *40*, 115-124.
5. Dufresne, A.; Cavaille, J. Y.; Vignon, M. R. *Journal of Applied Polymer Science* 1997, *64*, 1185-1194.
6. Dufresne, A.; Cavaille, J. Y. *J. Polym. Sci. Pt. B-Polym. Phys.* 1998, *36*, 2211-2224.
7. Dufresne, A.; Dupeyre, D.; Vignon, M. R. *Journal of Applied Polymer Science* 2000, *76*, 2080-2092.
8. Dufresne, A.; Cavaille, J. Y.; Helbert, W. *Polymer Composites* 1997, *18*, 198-210.
9. Helbert, W.; Cavaille, J. Y.; Dufresne, A. *Polymer Composites* 1996, *17*, 604-611.
10. Angles, M. N.; Dufresne, A. *Macromolecules* 2000, *33*, 8344-8353.
11. Chanzy, H.; Ernst, B.; Cavaille, J.-Y.; Favier, V. U.S. Patent 6103790, 2000.
12. Mathew, A. P.; Dufresne, A. *Biomacromolecules* 2002, *3*, 609-617.
13. Nair, K. G.; Dufresne, A. *Biomacromolecules* 2003, *4*, 657-665.
14. Berglund, L. In *Naturalfibers, biopolymers, and biocomposites;* Mohanty, A. K.; Misra, M.; Drzal, L. T., Ed.; CRC Press LLC, 2004; pp 807-832.
15. Zimmermann, T.; Pöhler, E.; Geiger, T. *Adv. Eng. Mater.* 2004, *6*, 754-761.
16. Zimmermann, T.; Pöhler, E.; Schwaller, P. *Adv. Eng. Mater.* in press
17. DIN EN 527-4, 1997.
18. Fischer-Cripps, A. *Nanoindentation;* Springer: New York, 2002.
19. Oliver, W. C.; Pharr, G. M. *Journal of Materials Research* 2004, *19*, 3-20.
20. Taniguchi, T.; Okamura, K. *Polymer International* 1998, *47*, 291-294.
21. Turbak, A. F.; Snyder, F. W.; Sandberg, K. R. *Journal of Applied Polymer Science. Applied Polymer Symposium* 1983, *37*, 815-827.
22. Herrick, F. W. U.S. Patent 4481077, 1984.
23. Hajji, P.; Cavaille, J. Y.; Favier, V.; Gauthier, C.; Vigier, G. *Polymer Composites* 1996, *17*, 612-619.
24. Dufresne, A.; Vignon, M. R. *Macromolecules* 1998, *31*, 2693-2696.
25. Samir, M.; Alloin, F.; Paillet, M.; Dufresne, A. *Macromolecules* 2004, *37*, 4313-4316.
26. Morin, A.; Dufresne, A. *Macromolecules* 2002, *35*, 2190-2199.
27. Dufresne, A. *Composite Interfaces* 2003, *10*, 369-387.

28. Favier, V.; Dendievel, R.; Canova, G.; Cavaille, J. Y.; Gilormini, P. *Acta Mater.* **1997**, *45*, 1557-1565.
29. Samir, M.; Alloin, F.; Sanchez, J. Y.; Dufresne, A. *Polymer* **2004**, *45*, 4 149-4157.
30. Favier, V.; Chanzy, H.; Cavaille, J. Y. *Macromolecules* **1995**, *28*, 6365-6367.
31. Ruiz, M. M.; Cavaille, J. Y.; Dufresne, A.; Gerard, J. F.; Graillat, C. *Composite Interfaces* **2000**, *7*, 117-131.
32. Matsumura, H.; Glasser, W. G. *Journal of Applied Polymer Science* **2000**, *78*, 2254-2261.
33. Hanley, S. J.; Gray, D. G. *Holzforschung* **1994**, *48*, 29-34.
34. Snell, R.; Groom, L. H.; Rials, T. G. *Holzforschung* **2001**, *55*, 511-520.
35. Chakraborty, A.; Sain, M.; Kortschot, M. *Holzforschung* **2005**, *59*, 102-107.
36. Xu, G. C.; Li, A. Y.; De Zhang, L.; Yu, X. Y.; Xie, T.; Wu, G. S. *Journal of Reinforced Plastics and Composites* **2004**, *23*, 1365-1372.
37. Roche, S.; Bec, S.; Loubet, J. L. *Mechanical Properties Derived from Nanostructuring Materials. Symposium. (Mater. Res. Soc. Symposium Proceedings Vol. 778)* **2003**, 117-122.
38. Stauss, S.; Schwaller, P.; Bucaille, J. L.; Rabe, R.; Rohr, L.; Michler, J.; Blank, E. *Microelectronic Engineering* **2003**, *67-8*, 818-825.
39. Borges, J. P.; Godinho, M. H.; Martins, A. F.; Stamatialis, D. F.; De Pinho, M. N.; Belgacem, M. N. *Polymer Composites* **2004**, *25*, 102-110.

Chapter 5

Morphology of Cellulose and Its Nanocomposites

B. S. Tanem[1], I. Kvien[1], A. T. J. van Helvoort[2], and K. Oksman[1]

Departments of [1]Engineering Design and Materials and [2]Physics, Norwegian University of Science and Technology, Trondheim, Norway

Atomic force microscopy and electron microscopy is demonstrated to give powerful insight into micro fibres of cellulose, cellulose whiskers and their nanocomposites with poly(lactic acid). Transmission electron microscopy, including bright field TEM and STEM appeared to be the most suitable way to examine cellulose whiskers and microfibers. AFM, however appeared to be the most suitable way at present to examine the bulk morphology of nanocomposites. The restricted resolution capabilities of scanning electron microscopy limits the information obtained. Scanning transmission electron microscopy combined with annular dark field detector is demonstrated to significantly increase the contrast of stained samples.

Introduction

Electron microscopy techniques like conventional bright field (BF) transmission electron microscopy (TEM) and scanning electron microscopy (SEM) have for decades been established as powerful techniques for detailed sub-μm morphology analysis in a wide range of materials, including soft materials. More recently, atomic force microscopy (AFM) was introduced (*1*) and has been applied successfully in a growing number of applications. Tapping mode AFM (*2*) opened the possibility for detailed examination of soft materials due to elimination of the lateral shear force, often damaging soft samples in contact mode AFM (*3*). The number of applications is growing fast, and AFM is reported to have some advantages compared to TEM. One of the principal advantages is the ability to perform imaging under liquids and to follow in-situ processes. In addition, AFM could go beyond the resolution that is easily achievable with TEM on soft materials. The soft materials have some principal difficulties in TEM due to electron-beam sensitivity and the need for staining to obtain sufficient contrast.

The potential of significant increase in mechanical-, thermal as well as barrier properties has triggered major research efforts in nanocomposites. In this growing field of nanotechnology the use of advanced characterization facilities with the sufficient resolution capabilities is an absolute requirement.

In this work the focus is towards the use of microscopy techniques to examine the reinforcement and bulk structures of nanocomposites where both the matrix material as well as the reinforcement is based on renewable resources. The high specific surface area together with the high modulus (*4*) makes microfibrillated cellulose (MFC) and cellulose whiskers to potentially attractive reinforcements, even for small loadings (*5*). The aspect ratio of MFC and the cellulose whiskers together with the dispersion and orientation in a matrix material will significantly influence the properties of the resulting nanocomposites. Hence, knowledge of these parameters is of fundamental importance which requires the use of microscopy techniques with sufficient resolution capabilities, often accompanied with labours sample preparation.

Conventional BF TEM has for decades been the preferred and routine methodology to examine the nanosized structures of various grades of MFC, see e.g. (*6*) and cellulose whiskers based on various sources, see e.g. (*7*). The information obtained is typically length, aspect ratio, shape and a qualitatively assessment of how successful the preceding disintegration process was. Metal shadowing or negative staining is commonly applied for contrast enhancement.

In addition, low-dose conditions are frequently applied to prevent or minimize beam damage. However, despite of its resolution capabilities, only a few works report the use of TEM in the study of bulk morphology in nanocomposite materials where nanocrystals based on cellulose are used as reinforcements. TEM examinations of nanocomposite films from dried suspension of tunicin whiskers mixed with non-structural polymer latices (8) or epoxy (9) have been reported. Sufficient contrast was reported to be present between the whiskers (6 wt%) and the latex matrix, without any need for staining (8). With epoxy as the matrix material and 5 wt% whiskers, OsO_4 was applied to obtain sufficient contrast (9). In addition, one work reports the use of TEM to examine quickly quenched suspensions of poly(caprolactone) and chitin nanocrystals by cryomicroscopy equipment (10).

Apart from the previous mentioned works, scanning electron microscopy (SEM) has almost exclusively been utilized for structure examination of nanocomposites using microfibrils (6) or various nanosized whiskers as reinforcements, see e.g. (7). The samples for SEM are typically frozen under liquid nitrogen, fractured and coated to avoid charging. SEM is reported to allow information about the dispersion and orientation of the whiskers in the matrix, in addition to the presence of aggregates and voids (7). In addition, SEM has also been utilized to examine microfibrils, see e.g. (11).

Limited work is present on the use of AFM to examine various cellulose microfibrils, whiskers and their nanocomposites. In an early work, Hanley et. al. (12) directly compare AFM and TEM images from the same area of microfibrils from V. ventricosa cell walls, pointing out the tip broadening effect in AFM and the influence of the scanning speed. This work and later work, see e.g. (13) has also performed high-resolution studies of the surface of microfibrils. The tip broadening effects has probably limited the use of AFM to examine nanosized cellulose whiskers. Two reports applying AFM on chitin nanosized whiskers are present (10,14), however, the images are not very clear. No reports are found using AFM to examine nanocomposites with MFC or cellulose whiskers as the reinforcement.

The aim of this work was to examine nanosized cellulose whiskers and MFC based on wood by several approaches including the use of field emission SEM, conventional BF TEM and AFM. In addition, more sophisticated TEM approaches are introduced and discussed. In addition, poly(lactic acid) – cellulose whisker nanocomposites are prepared and characterized by various approaches. Since PLA is a biodegradable polymer which could be processed like polyolefins, it is of particular interest as a matrix material.

Experimental

Materials

The PLA, Nature Works[TM] 4031D, was kindly supplied by Cargill Dow (USA). The cellulose whiskers were generated from micro crystalline cellulose by sulphuric acid treatment as described elsewhere (*15*). The micro fibers of cellulose (MFC) was obtained from refining and cryocrushing of northern black spruce bleached kraft pulp, described elsewhere (*16*). Nanocomposite films of PLA and cellulose whiskers were prepared by solution casting, as described elsewhere (*17*).

Atomic Force Microscopy (AFM)

For AFM analysis of cellulose whiskers, a droplet of the aqueous whisker suspensions was allowed to dry on a copper grid covered with a holey carbon film. This is convenient, since the same grid can be loaded into a TEM and thereby the same area could be examined by both AFM and TEM. AFM and TEM examination from the same areas of MFC has been demonstrated elsewhere (*12*). Therefore, for AFM analysis of MFC suspensions, a droplet of the aqueous suspension was allowed to dry on a freshly cleaved mica surface. Mica was chosen as the substrate since it is extremely smooth. For the bulk structure analyses, rectangular sheets were carefully cut out from the solution cast films, embedded in epoxy and allowed to cure overnight. The samples were thereafter trimmed in a Reichert-Jung ultramicrotome with freshly cleaved glass knives at -75°G to obtain a rectangular block surface, 50 x 500 μm^2 in cross-section. The final cuffing was performed with a diamond knife using a cuffing speed of 0.4 mm/s generating foils of ~50 nm in thickness.

The AFM measurements were performed with a NanoScope IIIa, Multimode[TM] SPM from Veeco. Calibration was performed by scanning a calibration grid with precisely known dimensions. All scans were performed in air with commercial Si Nanoprobes[TM] SPM Tips for Tapping mode. Height- and phase imaging were performed simultaneously in Tapping mode at the fundamental resonance frequency of the cantilever with a scan rate of 0.5 line/s using a j-type scanner.

Electron Microscopy

The whiskers and microfibers for SEM and TEM were prepared equally. A droplet of the diluted suspensions of cellulose whiskers/MFC was allowed to float on a copper grid and eventually flow through the copper grid covered with a holey carbon film. The samples were thereafter stained by allowing the grids to float in a 2 wt% solution of uranyl acetate for 3 min. For SEM analyses the grids with cellulose whiskers were mounted in a specialized holder, originally designed for scanning TEM to minimize background signal.

Bulk samples for TEM were prepared similar as for AFM, however, for TEM the microtomed foils were gathered onto cupper grids. The foils were thereafter stained by allowing the grids to float in a 2 wt% solution of uranyl acetate for 3 min.

All SEM analyses were performed in a Hitachi 4300S Field emission SEM (FESEM) at various accelerating voltages. Samples for TEM and scanning TEM were examined in a Philips GM 30 TEM at 150kV and a JEOL 2010 field emission TEM at 200 kV.

Results and Discussion

Whiskers and Micro Fibres of Cellulose

FESEM was found to be a quick way to get an overview of the dried cellulose whisker suspensions (Figure 1). The heavy element staining enabled sufficient Z-contrast for back-scattered electron imaging. Large amounts of cellulose whiskers were observed on the porous carbon film. The whiskers were not evenly distributed over the carbon film but occurred densely in some areas and were almost completely missing in others. This does not necessarily reflect the state of the suspension, but could rather be a result of the drying. At higher magnifications the presence of whiskers and agglomerates was more evident, however the resolution was considered insufficient for detailed information. The presence and shapes of the whiskers were defined through the heavy element staining surrounding the whiskers. Detailed measurements of the dimensions of the whiskers were therefore difficult to perform, also on images based on secondary electrons having slightly better resolution. However, most of the structures observed, believed to represent individual whiskers were in the range from 200 to 600 nm long. The diameter of the whiskers could not be measured from the images.

Conventional BF TEM analysis of the dried suspensions gave a clear overview of the dried whisker suspensions, and could also demonstrate the presence of larger fragments left after the preceding hydrolysis (Figure 2). The uranyl acetate staining gave reasonable contrast between the whiskers and the carbon film. The whiskers did not differ significantly in contrast from the carbon film, however the presence of the heavy uranium in close vicinity of each whisker gave enough contrast for imaging. Especially in those areas where the concentration of whiskers was rather high, the staining tended to form a continuous background, giving good contrast between the whiskers and carbon film. The length of individual whiskers was somewhat shorter than measured from FESEM, i.e. in the range from 100–300 nm. This difference is probably due to uncertainties in determining the actual length of the whiskers in the FESEM, since the staining partly shields the whiskers and the lack of resolution. The majority of the whiskers were found to have a diameter in the range from 10-15 nm. TEM also revealed that some of the whiskers were thicker at the middle than at the ends, however it is not clear whether this was an effect of agglomeration or the preceding hydrolysis.

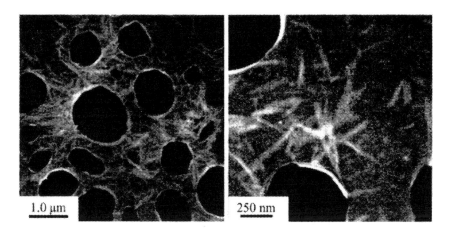

Figure 1. FESEM backscatter images of cellulose whiskers dried onto a holey carbon film at low magnification (left) and high magnification (right).

However, as will be further commented later in this text, even though conventional BF TEM appeared to give sufficient contrast between the cellulose whiskers and the carbon foil on dried cellulose whisker suspensions, the contrast is in general observed to be insufficient for bulk morphology analysis of nanocomposites of poly(lactic acid) (PLA) and such whiskers. This is in particular the case for low loadings. It is therefore of interest to explore possible ways to increase the contrast between the whiskers and a carbon rich matrix.

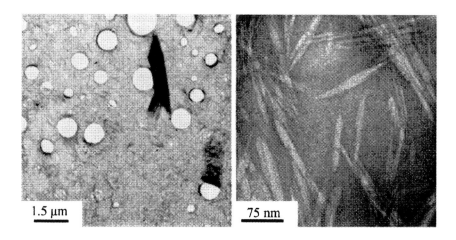

Figure 2. Conventional bright field TEM images of cellulose whiskers dried onto a holey carbon film at low magnification (left) and high magnification (right).

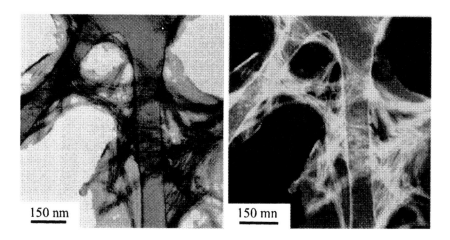

Figure 3. STEM ABF (left) and STEM ADF (right) of cellulose whiskers dried onto a holey carbon film.

Annular bright field (ABF) and annular dark field (ADF) scanning TEM (STEM) have some interesting advantages over conventional BF TEM. A high angle ADF (HAADF) image is formed by incoherently scattered electrons and could therefore utilize to a much larger extent the Z-difference introduced by the heavy element staining. By applying ABF and ADF STEM, with a well-defined 1 nm analytical probe, the contrast was observed to increase significantly (Figure 3). Work is in progress to utilize STEM on nanocomposites.

AFM tapping mode of the dried cellulose whisker suspensions (Figure 4) showed to be a reasonable alternative to electron microscopy, avoiding the limitations due to low contrast and resolution. Good contrast between the carbon foil and the whiskers were obtained in both height- and phase images. The length of the whiskers correlated with the length measured from TEM. Furthermore, the diameter of the whiskers was measured to be in the range from 20-30 nm, which is in general somewhat larger than observed from TEM measurements. The higher values of the diameter measured from AFM images was probably an effect of tip broadening, as reported previously on MFC (*12*). The AFM scanning of the cellulose whiskers was performed directly on a TEM grid. However, a direct comparison between AFM and TEM from exactly the same area was challenging, since typical areas that were suitable for AFM scans (e.g. Figure 4) did not give sufficient contrast for TEM after staining. In addition, areas being sufficiently stained for TEM were almost unsuitable for AFM analysis, since the uranyl acetate film close to the whiskers disrupted the AFM phase contrast images. A STEM ABF image at higher magnification (using a 0.2 nm probe) from an area close to the area shown in the AFM image in Figure 4, shows the presence of whiskers with some variation in appearance and thickness. STEM ABF allowed for quite accurate diameter measurements, with the most common thickness in the range from 10-15 nm, as indicated in the image. This is in good agreement to the BF TEM measurements.

A 20 nm line scan across a whisker at high magnification is shown in Figure 5. The whisker had a diameter of 6 nm. The image also illustrates that the uranyl acetate staining partly covers the whiskers. This could limit accurate measurements by other TEM techniques due to their lower resolution and lower contrast. These techniques could therefore have problems to properly define the interface between the whiskers and the matrix.

The MFC from refining and cryocrushing of northern black spruce bleached kraft pulp MFC was in general observed to be easier to characterize, since the major part of the material had dimensions that were somewhat larger than the whiskers. However, since the microfibres could be several μm long, it is difficult to get a proper length estimate, partly due to entanglements and problems to identify both ends. FESEM failed to identify properly the thinnest microfibrils. Typical observations based on conventional BF TEM observations at two magnifications are shown in Figure 6. The low magnification image shows the typical entanglements. It is also clear that this sample contained large

56

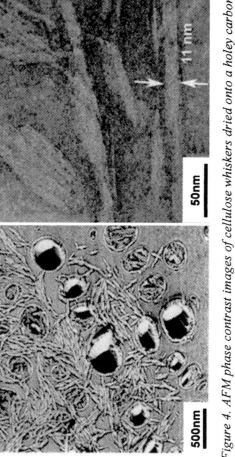

Figure 4. AFM phase contrast images of cellulose whiskers dried onto a holey carbon film at low (left) and STEM ABF image of whiskers from a nearby area (right).

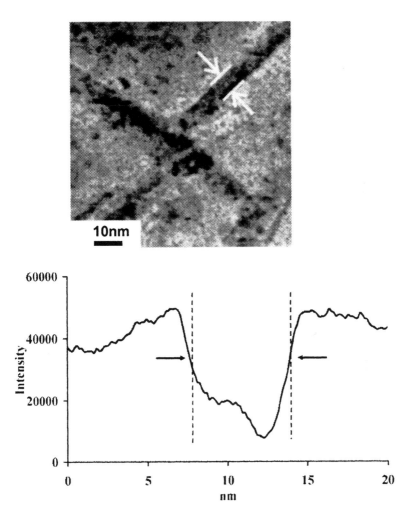

Figure 5. High magnification STEM HAADF of cellulose whiskers (top) and the corresponding line scan across one whisker (bottom).

58

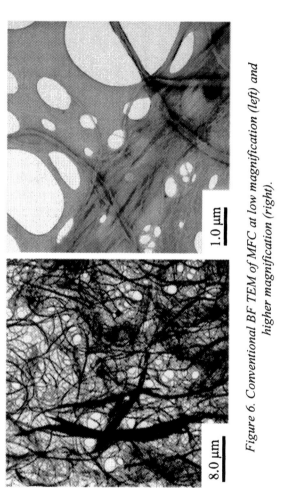

Figure 6. Conventional BF TEM of MFC at low magnifcation (left) and higher magnifcation (right).

fibre fragments, in addition to micro fibres and microfibrils. The image at higher magnification also illustrates this and shows in addition the splitting of a fibre fragment into micro fibres and the further splitting into microfibrils. The diameter of the thinnest microfibrils was difficult to observe due to entanglements, however, sizes down to ~ 10 nm were identified.

AFM analysis of MFC was done on a freshly cleaved mica surface instead of holey carbon film on a cupper grid (Figure 7).

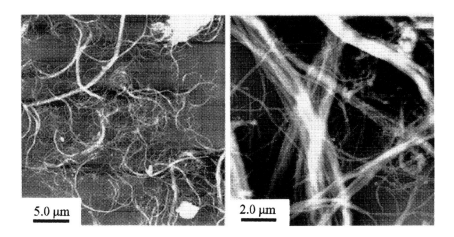

Figure 7. AFM tapping mode height images of MFC dried onto a mica surface at low magnification (left) and higher magnification (right).

Again AFM tapping mode turned out to be a good alternative to TEM i.e. details observed in TEM could easily be observed also in AFM, without the need for staining. However, as for the whiskers, AFM seemed to overestimate the diameter of the fibrils, due to the tip broadening effect. The thinnest fragments were observed to be ~25 nm in diameter.

Cellulose Whiskers Nanocomposites

Conventional BF TEM bulk analysis of the cellulose whisker nanocomposite was challenging for several reasons. The major problem was lack of contrast between the whiskers and the PLA matrix. Also the stability of the microtomed foils under the electron beam was insufficient, probably due to heating and/or charging effects. In local areas where the staining appeared to be more concentrated, probably due to higher amounts of whiskers, some insight into the structure was obtained (Figure 8). The staining in these areas made up a continuous background, giving reasonable contrast between the whiskers and the surroundings. In areas where the whiskers were more evenly dispersed, the contrast between the whiskers and the PLA matrix was low and insufficient for imaging. TEM analysis allowed for determination of the whisker length in the matrix. It seemed that the size of the whiskers was of the same range as before processing, i.e. the solution casting process did not affect the whiskers geometry.

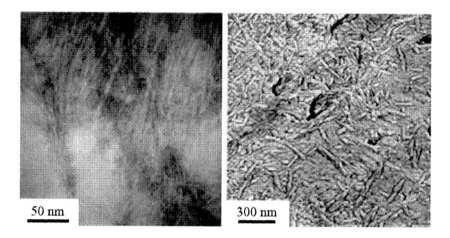

Figure 8. Conventional BF TEM image (left) and AFM phase contrast image (right) of bulk morphology in a solution cast PLA-cellulose whisker nanocomposite.

As previously discussed, STEM ABF and STEM ADF showed a significant increase in contrast between the carbon matrix and the whiskers and could therefore be a powerful way to study these nanocomposites. STEM also have a lower electron dose and a better control of the contrast and brightness of the signal and image compared to a conventional TEM. Nevertheless, more work is needed to sufficiently stabilize the microtomed foils during examination. This includes the use of a cooling stage down to cryogenic temperatures, i.e. the samples are examined at - 160°C. This will significantly reduce instabilities due to local heating and reduce carbon contamination.

FESEM examination of the surfaces of cryofractured samples gave no clear indications of the presence of whiskers. Both coated and uncoated surfaces were examined. The uncoated surfaces were examined at low accelerating voltages (1kV) to reduce charging effects. Since no heavy elements were introduced into the samples for FESEM analysis, any direct or indirect observation of whiskers in the bulk morphology must be related with characteristic topography of the fractured surfaces, resembling the typical shape and dimensions of the whiskers. No such features were clearly identified.

AFM analysis of surfaces generated by cryomicrotomy enabled detailed information of the bulk morphology of the solution cast PLA-cellulose whisker nanocomposites (Figure 8). The whiskers were clearly discerned from the matrix in both topography (not shown here) - and phase images. The whiskers partly protruded from the PLA matrix, giving rise to the contrast in the topography images. This effect is most probably due to differences in thermal expansion during heating from cryogenic temperatures. The contrast in the phase images appeared from the interphase between the whiskers and the matrix, probably a scanning effect. AFM therefore has the necessary resolution without the need for staining to increase contrast and without needing any further stabilization. AFM also enabled examination of significantly larger areas than TEM, since most of the areas in TEM lacked the necessary contrast between the whiskers and the matrix. Nevertheless, the use of more defined AFM probes would be beneficial to decrease the tip broadening effects.

Conclusions

The aim of this work was to explore and partly compare the ability of FESEM, AFM and different TEM approaches to examine micro fibres and whiskers of cellulose and cellulose whiskers nanocomposites.

Field emission scanning electron microscopy (FESEM) allowed for a quick examination giving an overview of the whiskers and microfibers of cellulose, however, the resolution was considered insufficient for detailed information. AFM suffered somewhat from the tip broadening effect, but do not require additional staining. TEM, including conventional BF TEM, ABF STEM and in particular ADF STEM appeared to be the most reliable approach to examine the dimensions of the whiskers and microfibres.

Ultramicrotomy of nanocomposite films at cryogenic temperatures enabled detailed inspection of the cellulose whiskers in the poly(lactic acid) matrix by AFM. Much larger regions could be examined by AFM compared to TEM, since the contrast between the whiskers and the matrix was insufficient in most areas in TEM. The analysis was carried out at ambient conditions without need for any pretreatment of the sample, in contrast to electron microscopy. AFM could therefore be a powerful alternative to TEM in such composite materials where contrast between the whiskers and the matrix is limited and the beam sensitivity is a major challenge. However, STEM ABF and STEM ADF in particular, have potential to significantly increase the contrast, but more work is needed to sufficiently stabilize the sample. FESEM applied on fractured surfaces allowed no insight into the detailed morphology of the nanocomposite.

Acknowledgements

Cargill Dow LLG, Minnetonka, USA is acknowledged for the supplied PLA polymer (Nature Works™). Professor M. Sain, University of Toronto, Canada is acknowledged for the micro fibrillated cellulose and Mr. D. Bondeson at our department for the prepared cellulose nano whiskers. The Norwegian Research Council is acknowledged for partial financial support of this work under the NANOMAT program.

References

1. Binnig, G.; Quate, C. F.; Gerber, C. *Phys. Rev. Lett.* **1986**, *56*, 930.
2. Zhong, Q.; Inniss, D.; Kjoller, K.; Elings, V. B. *Surf Sci.* **1993**, *290*, L668.
3. Maivald, P.; Butt, H. -J.; Gould, S. A. C.; Prater, C. B.; Drake, B.; Gurley, J.; Elings, V. B.; Hansma, P. K. *Nanotechnology* **1991**, *2*, 103.
4. Tashiro, K.; Kobayashi, M. *Polymer* **1991**, *32*, 1516.
5. Felix, J. M.; Gatenholm P. *J. Appl. Polym. Sci.* **1991**, *42*, 609.
6. Malainine, M. E.; Mahrouz, M.; Dufresne, A. *Compos. Sci. Technol.* **2005**, *65*, 1520.
7. Samir, M. A. S. A.; Alloin, F.; Dufresne, A. *Biomacromolecules* **2005**, *6*, 612.
8. Favier, V.; Chanzy, H.; Cavaillé, J. Y. *Macromolecules* **1995**, *28*, 6365.
9. Ruiz, M. M.; Cavaillé, J. Y.; Dufresne, A.; Graillat, C.; Gérard, J- F. *Macromol. Symp.* **2001**, *169*, 211
10. Morin, A.; Dufresne, A. *Macromolecules* **2002**, *35*, 2190.
11. Zimmermann, T.; Pöhler, F.; Geiger, T. *Adv. Eng. Mater.* **2004**, *6*, 754.
12. Hanley, S. J.; Giasson, J.; Revol, J-F.; Gray, D. G. *Polymer* **1992**, *33*, 4639
13. Baker, A. A.; Helbert, W.; Sugiyama, J.; Miles, M. J. *Biophys. J.* **2000**, *79*, 1139.
14. Lu, Y.; Weng, L.; Zhang, L. *Biomacromolecules* **2004**, *5*, 1046.
15. Bondeson, D.; Mathew, A.P.; Oksman, K. *Cellulose,* submitted.
16. Chacraborty, A.; Sain, M.; Korschot, M. *Holzforschung* **2005**, *59*, 102
17. Kvien, I.; Tanem, B. S.; Oksman, K. *Biomacromolecules* **2005**, *6*, (*6*), 3160.

Chapter 6

Useful Insights into Cellulose Nanocomposites Using Raman Spectroscopy

S. J. Eichhorn

Materials Science Centre, School of Materials, Grosvenor Street, Manchester M1 7HS, United Kingdom

With the drive to move towards renewable resources, and with the fundamental scientific desire to better understand the materials that nature has provided us with, in terms of their mechanics and hierarchical structure, the study of cellulose-based nanocomposite materials has increased dramatically in the last 10 years. This chapter introduces the concepts of using Raman spectroscopy for the study of cellulose nanocomposites. The principles behind Raman spectroscopy are described, with particular emphasis on the information it can yield. The shift in Raman bands towards a lower wavenumber upon the application of external deformation is described as a means of not only following molecular deformation, as has been reported for a wide variety of other materials, but also as a tool for monitoring the dispersion of the nanostructured phase and therefore the level of stress-transfer.

Introduction

It has been established from early studies on needle-like fibres, that high strengths can be obtained if one reduces the diameters of such elements (*1*). Since the pioneering work of Frey-Wyssling (*2*) on what we would now term the nanostructure of cellulose and the work of the late Reginald D. Preston FRS *(3)* on accurate measurement of these "microfibrils, a term that was objected to when it was originally coined (*4*), there have been many studies to better understand their mechanical properties. Subsequently, it has become clear that their potential use in composites could be realized.

The use of cellulose nanofibres in composite materials probably has its origins in the use of nanofibrillated cellulose for the reinforcing of paper-sheets, as was reported some time ago in the early 1980's (*5*). This is however a concept that has been known for some time since small sub-micron fibrils of cellulose can be released from the cell-walls of fibres through mechanical refining to enhance bonding *via* an increase in the available surface area (*6*).

Other forms of nanostructured cellulose for composite and paper reinforcement have been investigated, such as bacterial cellulose (*7,8*) and more recently the use of xyloglucan endotransglycosylase (XET) for the glycosilisation of xyloglucan chains to adhere cellulose fibres to resin matrices (*9*). It is however the work of Dufresne *et al.* (*10-15*) that has spearheaded much research into nanostructured cellulose composites, although others are noted for their much earlier contributions in producing nanocrystals (*16*), and in the area of mimicking the plant cell wall with such materials (*17*).

However, despite all these advances it is fair to say that little is known about the details of the interface between these materials and the matrices that they are incorporated into to reinforce. Little is also known about the mechanical properties of these materials, as it is very difficult to test small fibres of small dimensions effectively using conventional means. There have been however some recent studies using AFM (Atomic Force Microscopy) (*18*) and Raman spectroscopy (*19*) that have been used to determine the stiffness of nanostructured cellulose in the bacterial and tunicate forms respectively. This article will report on the latter technique, and the insights that can be gained from using this to tell us about dispersion, stress-transfer and the fundamental properties of the material. Comparisons will be made with data obtained from other nanocomposites such as carbon nanotube based materials.

Raman Spectroscopic Studies of Deformed Polymers

Raman spectroscopy has been used to study the deformation micromechanics of a large number of polymeric fibres, and in more recent times, nanostructured materials such as carbon nanotubes and others. The first published observation of a Raman band shift in a polymeric material was that by Mitra *et al.* in 1977 (*20*) on polydiacetylene single crystal fibres. These showed quite clearly that if stress was applied to the single crystals then clear shifts in a number of peaks, corresponding to main chain moieties such as -C ≡C- and -C=C-, were observed. This observation was explained theoretically (*20*) and subsequently in more detail by Batchelder and Bloor (*21*). A formalized relationship between the vibrational frequency (expressed in this case as reciprocal wavenumbers) of a band and an anharmonic force constant coefficient (F_2) has been given by Mitra *et al.* (*20*) as

$$\upsilon = \left(4\pi^2 c^2 M^* \right)^{-1/2} \left(F_2' \right)^{1/2} \tag{1}$$

where c is the speed of light and F_2' is the second force constant coefficient in the expansion of the second derivative of the potential energy system of N anharmonically oscillating atoms.

In terms of cellulose, the first shifts in their Raman spectra were first observed in 1997 for regenerated cellulose fibres (*22*), and subsequently in a large number of other fibres (*23,24*) including natural fibre samples such as flax and hemp (*25,26*). This shift has also been recently confirmed by independent groups on fibres (*27*) and also in composite materials (*28*). The technique has been used quite extensively to map stresses in cellulose-based composite materials (*29,30*) and fibrous networks (*29*). To this end, the fibre molecular deformation, as revealed by a shift in the position of the Raman peaks with applied tensile deformation, can be used as gauge of local deformation in the composite.

It is possible to theoretically determine the shift in Raman band using computer modelling techniques. However, a simple approach, as employed by Mitra *et al* (*20*) based on Badger's law (*31*) which relates the force constant *(K)* of a group of atoms to their separation *(Rₓ)* by the empirical equation

$$K^{1/3} = a \left(R_x - b \right) \tag{2}$$

where a and b are fitting constants and have been derived for a large number of periodical groups of atoms (31). Equations (1) and (2) can be combined, and by manipulation one can obtain the approximate relationship (20)

$$\Delta \upsilon \cong -\frac{3}{2}\frac{\upsilon}{2(R_x - b)}\Delta R \qquad (3)$$

Now for cellulose, and for the widely reported (22-28) shift in the 1095 cm^{-1} band for the C-O stretch mode corresponding to the ring stretching modes of cellulose (32) and possibly the glyosidic ring stretch (33), it is possible therefore to predict the magnitude of the shift in this band $(\Delta \upsilon)$. Taking values of b from Badger (31) to be 0.68 Å and the bond length for the C-O moiety from Treloar (36) to be 1.43 Å one can obtain a value of -3 cm^{-1} for $\Delta \upsilon$ assuming that the bond strain is only 10% of that applied to the polymer (say 1%). We will return to this value at a later date when discussing the results obtained during experiments on nanostructured cellulose composites.

A more rigorous approach is to take an empirical forcefield and minimize a series of atoms within this field under restraint using a computer simulation package. A normal mode analysis can then be performed on this group of atoms. Normal mode analysis is applied to the well- minimize system to determine theoretical spectral intensities (analogous to infrared frequencies). They are usually calculated by the software from a static situation using second derivatives of the potential energies. One can write the kinetic energy (T) of the a general system of N atoms with mass m_i as

$$T = \frac{1}{2}\sum_{i=1}^{3N} m_i \left(\frac{dq_i}{dt}\right)^2 \qquad (4)$$

where q_i is a general displacement of the system.

This can be expanded as a Taylor series yielding the potential energy of the system as

$$U = U_0 + \sum_{i=1}^{3N}\left(\frac{\partial U}{\partial q_i}\right)_0 q_i + \frac{1}{2}\sum_{i,j=1}^{3N}\left(\frac{\partial^2 U}{\partial q_i \partial q_j}\right)_0 q_i q_j + \ldots \qquad (5)$$

If the system is well minimized then U_0 will be close to zero, and therefore the potential gradient ought also to be zero. Therefore the second term disappears in the expansion. Since we are also using a harmonic approximation at this point, higher terms in the expansion can also be ignored. If one substitutes these values in Newton's equations of motion, one gets sets of coupled second-order differential equations of the form

$$\frac{d^2 q_i}{dt^2} + \sum_{j=1}^{3N} \left(\frac{\partial^2 U}{\partial q_i \partial q_j} \right)_0 q_j = 0, \quad i = 1, 2 \dots 3N \tag{6}$$

Such equations have a solution of the form

$$q_i = A_i \cos(\omega t + \varphi) \tag{7}$$

where A_i is a constant (amplitude), ω is a frequency term and φ represents a phase change. When these solutions are substituted into equation 6 the resultant simultaneous equations can be written as a single matrix eigenvalue equation

$$\overline{M} = \ddot{\underline{q}} = \omega^2 \underline{q} \tag{8}$$

where $\overline{M}$ is a matrix of second derivatives of U, $-\omega^2$) are the eigenvalues and q are the eigenvectors. Frequencies can be obtained by creating a matrix of mass weighted second derivatives and by diagonalising one obtains the eigenvalues. Substituting these values back into the matrix equation, and solving the resultant simultaneous equations obtains the eigenvectors, or the directions in which the vibrations are acting.

Similar normal mode analyses, using a variety of commercial and dedicated forcefields have been used to model the molecular deformation of polymers, as revealed by Raman spectroscopy of fibres under deformation. In recent times however, it has become possible to determine theoretical Raman band shifts in cellulosic structures using the approach as described above (19,34) and this article will review such approaches and how it could be extended to better understand the fundamental deformation of nanostructured material. In recent times (35) it has also been possible to construct theoretical models of a composite of a nanostructured cellulose material with a resin interface, and to Determine for the first time properties of the interface. This approach offers a

68

unique insight into interfacial properties, and if nothing else gives a benchmark of a perfect interface against which all of our laboratory and industrially manufactured materials can be compared to.

Raman Spectroscopic Studies of Deformed Composites

As has already been described, the shift in a particular Raman band can be used to determine the local molecular deformation of cellulose fibres deformed in tension. An example of a typical shift in a natural fibre (reproduced from (*30*) is shown in Figure 1.

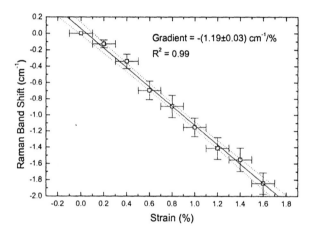

Figure 1. A typical Raman band shift for a cellulose fiber (hemp) with applied strain (Reproduced from reference 30. Copyright 2004 Elsevier)

It is clear that discernable Raman band shifts can be obtained from such material, and as we shall see later it is possible to relate the shifts observed in nanostructured cellulose to obtain their mechanical properties.

Before embarking on a review of how Raman spectroscopy can and could be utilized for the study of cellulose nanocomposites, it is worth mentioning the various forms of model composites that have been traditionally used for the study of interfacial properties. Model composites, usually of a single fibre and much larger volume of matrix material, are employed because this enables the mechanical properties of the fibre-matrix interface to be isolated without the influence of the clustering of fibres in a high volume fraction material. It should be stressed however, that these types of composite are not generally achieved with nanocomposites, since it is difficult to isolate and manipulate single fragments of such material without recourse to damage or difficulties in separation. The main forms of model composites are as follows; single fibre pull-out, push-in, fragmentation and microbond. In recent times, a new form of geometry in the form of a microdroplet test, has been reported (28,30). These geometries are schematically reported in Figure 2, and a review of the stress-transfer properties of these using Raman spectroscopy has been reported by Young (37). These model composites, as already mentioned would be difficult to reproduce on the nanoscale, although work has recently been reported by Wagner *et al.* (38) on a form of pull-out test performed on carbon nanotube composites using an AFM (Atomic Force Microscope) tip. The reinforcing of composites using nanostructured components, such as cellulose for instance, is hampered by the fact that adequate dispersion of the reinforcing components is not possible. This also negates the possibility of truly understanding the interface between single elements and the matrix, and recourse to theoretical and often quite empirical relationships of the reinforcement are utilized to explain phenomena where clustering takes place.

In terms of the structure/property relationships of cellulose nanocomposites, as elucidated using Raman spectroscopy, it is important also to discuss the specimen geometries that have been used to study these materials. A 4-point bending experiment has recently been conducted on both tunicate and sugarbeet cellulose composites wherein samples were dispersed in polymeric resin which itself was secured to a thin beam of the same material, in this case epoxy resin, and deformed under the microscope of the Raman spectrometer (19) in tension. In this geometry, and given the spot size of the Raman spectrometer used (about 2 μm), it is neither possible to determine stress on the sample (only strain via a gauge) or the properties of single fragments of material, as they have dimensions in the nanometer range. On the latter point it is therefore only possible to average over a large number of fragments of cellulose, and by using assumptions about the stress-transfer and the formation of the composites, mechanical properties can be established. Before summarizing these results, and the insights obtainable by these methods, the theoretical background and the rationale to this approach will now be dealt with in depth.

If one considers that you have a 2-D random dispersion of fibres in the resin, then one can use a form of analysis developed by Krenchel (39) to obtain the mechanical stiffness of a single fragment. To establish that there is indeed a 2-D random dispersion one can use the fact that the laser source can be polarized in one direction; either parallel or perpendicular to the longest edge of the sample. In this respect it is possible to obtain a maximum intensity Raman signal from fragments only oriented in the polarization direction, and by assuming that the cellulose chains in the fragments are perfectly oriented then one can use this to obtain an intensity *versus* rotation angle by rotating the specimen with respect to the polarization direction. The recorded intensity, of a main chain Raman vibration, such as the 1095 cm^{-1} peak for instance, is directly related to the

70

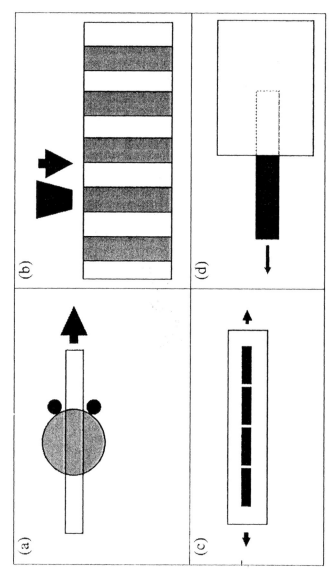

Figure 2. Schematics of the common composite interfacial tests, (a) the microbond (b) the push-in (c) the fragmentation and (d) the pull-out tests.

orientation of the fibres in that polarization direction, and so a polar plot can be obtained. An example of such a plot is shown in Figure 3 for a tunicate sample where it can be seen that there is no preferential orientation of the material, and the assumption of random dispersion can be made.

Krenchel analysis was developed in the early 1960's (39), the basic premise of which is that the reinforcement provided by fibres orientated randomly within a composite is governed by an efficiency factor. These efficiency factors (no) can be used to determine the modulus of a composite (E_c) given the modulus of a fragment of the material (E_m) via the equation

$$E_c = \eta_0 E_m \qquad (9)$$

These efficiency factors have been analytically determined (39, 40) as $9/8\pi$ for 2-D in plane random and for 3-D random 1/5. In this respect it is possible to back-calculate the stiffness of a single fragment of nanostructured cellulose, provided that one assumes that a random dispesion of the elements is obtained. It is also noted that this only gives an estimate, and other factors such as transverse properties and the contribution of off-axis fibres to the stiffness are not incorporated, but nevertheless could be with a more in depth approach. Typical shifts in the position of the 1095 cm^{-1} Raman band for tunicate are shown in Figure 4 and a shift in the same peak for sugarbeet is shown in Figure 5. It is clear that the tunicate samples show characteristic behaviour, wherein the band shifts by a large amount initially after which there is a plateau in this shift at about 0.8 % strain. The initial shift is thought to be due to direct deformation of the molecular backbone of cellulose, as has been previously postulated (22-30) and the onset of the plateau is thought to be due to debonding of the cellulose fibres from the matrix. This debonding was not found to occur with the sugarbeet cellulose, and the gradient of a linear fit through the data points is smaller in magnitude for this sample than observed with the tunicate. The latter was found (19) to have a maximum slope of -2.5 cm^{-1}/% which is close to the value of -3 cm^{-1}/% predicted by calculation in the previous section using Badger's law (31) and an approach by Mitra et al. (20). The magnitude of this parameter is thought to be indicative of the stress-transfer properties of the composite, wherein a higher value may represent for instance better dispersion of the nanostructured component or a better interface between resin and fiber. However, by the very fact that this is close to the theoretical value for a C-O bond suggests that the transfer of stress to the cellulose backbone is indeed very good. This value of -2.5 cm^{-1}/% is much greater than has been obtained for a large number of cellulose fibres, as shown in Figure 6, where the magnitude of the shift rate with respect to strain has been plotted for many samples against their moduli. It is clear from this diagram that the shift rate with respect to strain is proportional to the modulus of the fibers, and a correlation can be clearly made between these two parameters enabling an extrapolation of the predicted modulus of the nanostructured material under investigation here can be determined.

72

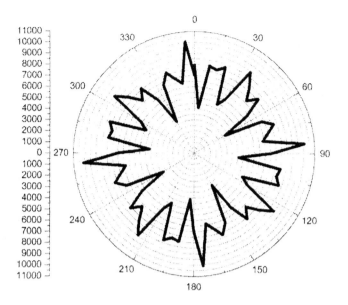

Figure 3. Polar plot of the intensity of the 1095 cm⁻¹ peak for a tunicate-epoxy composite as a function of rotation angle (measurements from 0-90° are mapped into the remaining 3 quadrants). The intensity is reported on an arbitrary scale on the left and the solid line is for the eye only (individual values were taken at 5° intervals)

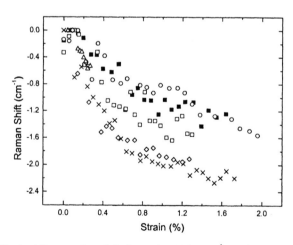

Figure 4. Typical Raman band shifts in the 1095 cm⁻¹ peak for tunicate-epoxy composites with strain (deformed in 4-point bending)

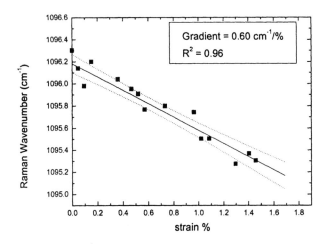

Figure 5. A typical Raman band shift for a sugarbeet cellulose composite deformed in tension using a 4-point bend test.

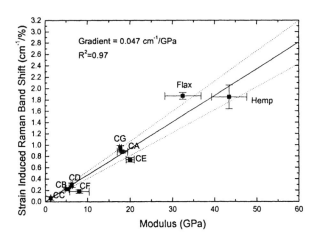

Figure 6. The strain induced Raman band shift rate versus the modulus for a number of regenerated cellulose (CA-CG), flax and hemp fibers.

74

If this is done then values of 51 and 13 GPa are obtained for the tunicate and sugarbeet moduli respectively. Now, this represents the modulus of the composite, or at least the 2-D random dispersion of fibres, and therefore by using equation 9 and a value for the efficiency factor of $9/8\pi$ one can determine values for the stiffnesses of 143 and 36 GPa for the tunicate (*19*) and sugarbeet samples respectively. The first value is close to that obtained for cellulose-I fibres by Sakurada *et al.* (*41*) of 137 GPa. Using the theoretical computer simulation approach, as already outlined, it is possible to also predict the stiffness of a fully oriented chain of cellulose along with the Raman band shift that one would expect from such a structure. This has been done for cellulose-I (*19*) with a predicted stiffness of 143 GPa, which is close to that determined experimentally with the tunicate samples. This is an indication that the cellulose chains in the tunicate sample are highly oriented. However, the lower modulus of the sugarbeet sample indicates that these fibres either have a lower orientation, do not bind to the resin as effectively or that their crystallinity is not as high. A value of 36 GPa puts them mechanically in the same range as natural fibers such as highly oriented flax and hemp *(cf.* Fig. 6) which would suggest a lower crystallinity in these samples. However, this is a topic for future research and it is thought that the post-harvest processing of the fibres may also play a significant role in their properties. The predicted Raman band shifts for an oriented chain of cellulose-I (*19*) produced by normal mode-analysis are reported in Figure 7.

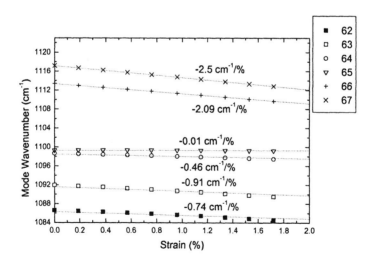

Figure 7. Predicted Raman band shifts for a fully oriented chain of cellulose-I determined using normal mode analysis (Reproduced from reference 20. Copyright 2005. American Chemical Society).

The band located at about 1116 cm^{-1} in Figure 9 has been (*19*) cited as representative of the experimentally observed vibration at 1095 cm^{-1}. The reason for this is because it was found to have a much lower infared intensity compared to the other bands predicted close-by. A low infrared intensity usually indicated that the Raman intensity will be conversely high, and hence this band was chosen for this reason. The gradient of the shift in this band with strain, as one can see from Figure 7 is the same magnitude as that observed in experiments on tunicate samples which is an indication of both the validity of the approach, and that the tunicate cellulose chains must be well-oriented.

Conclusions

It has been shown that Raman spectroscopy is a powerful technique for the monitoring of the deformation micromechanics of cellulose nanocomposites. It is possible to extract the mechanical properties of single fragments or fibres within the composite provided that one knows something about the dispersion. This is therefore still an indirect measurement of these properties, but nevertheless has yielded a value of 143 GPa for a highly oriented sample of tunicate which is of the same order of magnitude as that obtained both by theoretical means or from previous experiments. The relatively lower value obtained for the sugarbeet sample may be indicative of the lower orientation in this sample, less adhesion between fiber and matrix or the lower crystallinity of this material. However, the calculated stiffness (36 GPa) for sugarbeet is of the same order of magnitude as other native cellulose fibres such as flax and hemp and is an indication that it could be used in composite materials for reinforcement.

Acknowledgements

The author wishes to thank Dr. Adriana Stucova (Glasgow University) and Mr. Deepak Kalaskar for the work on the tunicate and sugarbeet cellulose composites. Professor J.-Y Cavaillé is gratefully acknowledged for supplying the tunicate and sugarbeet samples as is Professor G.R. Davies for his assistance in the modeling studies. Professor R.J. Young is also acknowledged for his help with the research. This work was completed using funding from the EPSRC under grant numbers GR/M82219 and GR/S44471.

References

1. Gordon, J.E. *Nature.* **1957**, 179, 1270.
2. Frey-Wyssling, A. *Papierfabrikant.* **1938**, 36, 212.
3. Preston, R.D. *Discussions of the Faraday Society.* **1951**, 11, 165.

76

4. *ibid,* 208.
5. Turbak, A.F.; Snyder, F.W.; Sandberg, K.R. *J. Appl. Polym. Sci: Appl. Polym. Symp.* **1983**, 37, 815.
6. Smook, G.A. *Handbook For Pulp and Paper Technologists,* 2nd Edition, Angus Wide Publications, 1992.
7. Tajima, K.; Fujiwara, M.; Takai, M.; Hayashi, J. *Mokuzai Gakkaishi.* **1995**, 41, 749.
8. Takai, M.; Nonomura, F.; Inukai, T.; Fujiwara, M.; Hayashi, J. *Sen'i Gakkaishi.* **1991**, 47, 119.
9. Brumer, H.A.; Zhou, Q.; Baumann, M.J.; Carlsson, K.; Teeri, T.T. *J. Am. Chem. Soc.* **2003**, 126, 5715.
10. Favier, V.; Canova, G.R.; Cavaillé, J.Y.; Chanzy, H.; Dufresne, A.; Gauthier, C. *Polym. Adv. Tech.,* **6**, 35.
11. Helbert, W.; Cavaillé, J.Y.; Dufresne, A. *Polym. Comp.* **1996**, 17, 604.
12. Dufresne, A.; Cavaillé, J.Y.; Helbert, W. *Polym. Comp.* **1997**, 18, 198.
13. Dufresne, A. *Composite Interfaces.* **2003**, 10, 369.
14. Samir, M.A.S.A.; Alloin, F.; Sanchez, J.-Y.; Dufresne, A. *Macromolecules.* 2004, 37, 4839.
15. Samir, M.A.S.A.; Alloin, F.; Sanchez, J.-Y.; Dufresne, A. *Polymer* **2004**, 45, 4149.
16. Marchessault, R.H.; Morehead, F.F.; Walter, N.M. *Nature.* **1959**, 184, 632.
17. Astley, O.M.; Chanilaud, E.; Donald, A.M.; Gidley, M.J. *Int. J. Biol. Macromol.* 2003, 32, 28.
18. Guhados, G.; Wan, W.; Hutter, J.L. *Langmuir.* **2005**, 21, 6642.
19. Sturcova, A.; Davies, G.R.; Eichhorn, S.J. *Biomacromolecules.* **2005**, 6, 1055.
20. Mitra, V.K.; Risen, W.M.; Baughman, R.H. *J. Chem. Phys.* **1977**, 66, 2731.
21. Batchelder, D.N.; Bloor, D. *J. Polym. Sci. B – Polym. Phys. Edn.* **1979**, 17, 569.
22. Hamad, W.Y.; Eichhorn, S. *ASME J. Eng. Mat. Tech.* **1997**, 119, 309.
23. Eichhorn, S.J.; Yeh, W.-Y.; Young, R.J. *Text. Res. J.* **2001**, 71, 121.
24. Eichhorn, S.J.; Young, R.J.; Davies, R.J.; Riekel, C. *Polymer.* **2003**, 44, 5901.
25. Eichhorn, S.J.; Hughes, M.; Snell, R.; Mott, L. *J. Mat. Sci. Lett.* **2000**, 19, 721.
26. Eichhorn, S.J.; Sirichaisit, J.; Young, R.J. *J. Mat. Sci.* **2001**, 36, 3129.
27. Fischer, S.; Schenzel, K.; Fischer, K.; Diepenbrock, W. *Macromol. Symp.* **2005**, 223, 41-56.
28. Tze, W.T.; O'Neill, S.C.; Gardner, D.G.; Tripp, C.P.; Shaler, S.M. *Abstracts of Papers of the American Chemical Society.* **2002,** 223, 080-CELL.
29. Eichhorn, S.J.; Young, R. *Comp. Sci. & Technol.* **2003**, 63, 1225.
30. Eichhorn, S.J.; Young, R.J. *Comp. Sci. & Technol.* **2004**, 64, 767.
31. Badger, R.M. *J. Chem. Phys.* **1934**, 2, 128.

32. Wiley, J.H.; Atalla, R.H. *Carbohydr. Res.* **1987**, 160, 113.
33. Edwards, H.G.M.; Farwell, D.W.; Webster, D. *Spectrochim. Acta. A.* **1997**, 53, 2383.
34. Eichhorn, S.J.; Young, R.J.; Davies, G.R. *Biomacromolecules* **2005**, 6, 507.
35. Chauve, G.; Heux, L.; Arouini, R.; Mazeau, K. *Biomacromolecules* **2005**, 6, 2025.
36. Treloar, L.R.G. *Polymer.* **1960**, 1, 290.
37. Young, R.J. *Key Eng. Mat.* **1996**, 116, 173.
38. Nuriel, S.; Katz, A.; Wagner, H.D. *Comp. A.* **2005**, 36, 33.
39. Krenchel, H. *Fibre Reinforcement.* Akademisk Forlag, Copenhagen, 1964.
40. Cooper, C.A.; Young, R.J.; Halsall, M. *Comp. A.* **2001**, 32, 401.
41. Sakurada, I.; Nukushina, Y.; Ito, T. *J. Polym. Sci.* **1962**, 57, 651.

Chapter 7

Novel Methods for Interfacial Modification of Cellulose-Reinforced Composites

Scott Renneckar[1], Audrey Zink-Sharp[1], Alan R. Esker[2], Richard K. Johnson[1], and Wolfgang G. Glasser[1]

Departments of [1]Wood Science and Forest Products and [2]Chemistry, Virginia Polytechnic Institute and State University, Blacksburg, VA 24061–0323

Cellulose surface modification is reviewed from the standpoint of the importance of interfacial modification for the development of strong cellulose reinforcement (nanowhiskers, wood fiber, or continuous fiber) in thermoplastic matrices. While continuous fiber composites did not require interfacial modification for improvement in mechanical properties, short fiber composites did. Three alternative methods for modification are highlighted: (a) esterification of pulp fibers in a nonswelling solvent produces a composite with covalent links between cellulose I crystals and a thermoplastic ester matrix; (b) reactive processing via steam-explosion of wood in the presence of polyolefins creates a wood fiber with a thermoplastic matrix surface; and (c) "bottom-up"-nanocomposites are created by the self assembly of irreversibly adsorbing amphiphilic polymers or polyelectrolytes onto cellulose surfaces, and alternatively by the layer-by-layer assembly of anionic cellulose nanowhiskers and cationic starch.

Introduction

Cellulosics are finding more opportunities as reinforcing agents and fillers in thermoplastic composites. Over the past decade, wood and natural fiber-filled thermoplastic composites have increased to a production level of 2 billion pounds per year (*1*). Part of the recent growth has been helped by research into coupling agents that addressed the performance of cellulosic/thermoplastic materials by compatibilizing the interface between the cellulose fiber and the matrix (*2*). Commercial coupling agents, available from at least six industrial suppliers, can increase mechanical properties from 20 to 200% (*3*).

There are many attributes that justify the inclusion of cellulose into composite materials. Mechanical properties of cellulose and cellulose derivatives can match synthetic counterparts with the added benefits of abundance and biodegradability. Yet, when compared with other materials cellulosics have lower thermal stability than many synthetic polymers, and dimensional stability is poor. However, there is still room for growth for cellulosic composites. Developments in solvents for cellulose, and isolation of nano-sized cellulose crystallites, "nanowhiskers", create opportunities for rapid processing of cellulosics and cellulosic-based materials with unique properties. For the former, n-methyl morpholine n-oxide is commercially used to spin cellulose from the liquid crystalline state, while for the latter, mineral acid hydrolysis is used to isolate native cellulose I crystallites. The crystalline cellulose structure derives its strength from inter- and intramolecular hydrogen bonds that provide for high mechanical properties with a modulus of 143 GPa (*4*), while strength can be estimated (on the basis of E/10 to E/5 (*5*)) between 14.3 GPa to 28.6 GPa. Hence, crystalline morphology increases the mechanical properties of the spun fiber, while isolated nanowhiskers approach the theoretical mechanical properties of the unit crystal. In short, crystalline cellulose can be considered a strong building block to be used within composite materials; however for it to serve as a reinforcement agent, stress needs to travel through the interface from the polymer matrix.

Strong cellulosic composites are realized by careful consideration of the interface and cellulose particle geometry. Interfacial interactions within cellulosic-thermoplastic composites govern the adhesion, water sorption, durability, and processing of the material. In this chapter, we review a variety of options for interfacial compatiblization for cellulose-reinforced composites, from continuous fiber to nanowhiskers. These options have included (a) partial chemical modification via esterification; (b) controlled interactions by reactive processing; and (c) a biomimetic approach of self-assembly with amphiphilic or ionic biopolymers in relation to a "bottom-up" design strategy. The latter refers to the engineering or multiphase materials by controlled molecular interactions that produce a hierarchical order in materials on many different dimensions. Aspects of the hierarchical organization of cellulose-lignin blends have recently been reviewed (*6*).

Cellulose Reinforced Composites:

Fiber reinforced composites can be classified into two categories: continuous fiber or discontinuous fiber reinforced polymer composites. For continuous fiber composites, mechanical properties are determined on the basis of the individual mechanical properties and volume fractions of fiber and matrix components (*7*). Glasser and co-workers (*8-10*) and Bourban et al. (*11*) produced strong composites from continuous regenerated cellulose (lyocell) fibers and various thermoplastic biobased polymer matrices. Cellulose acetate butyrate (CAB) composites with mechanical properties of 300 MPa for ultimate tensile strength and 20 GPa for modulus were produced with 60% lyocell fiber loading (*8*). Bourban et al. created a biodegradable composite from a 3-hydroxybutyrate-3-hydroxyvalerate (PHB/V) copolymer matrix (*11*). At 26.5% fiber volume fraction the composite material had an ultimate tensile strength and tensile modulus of 278 MPa and 11.4 GPa, respectively. Composites reinforced with a unidirectional continuous cellulose tow demonstrated that the rule of mixtures was an appropriate model for composite modulus (E_c), i.e. the modulus increased linearly as a function of fiber volume fraction (V_f) and modulus of the fiber (E_f) and matrix (E_m): ($E_c = (V_f E_f + V_{1-f} E_m)$, without any type of surface modification (Figure 1). Furthermore, after surface modification of the continuous fiber no statistically significant improvement in mechanical properties was detected (*9*). Based upon the rule of mixtures, fiber-matrix adhesion only contributes to the continuous composite material's mechanical properties when the material is tested in the transverse direction to the fiber (*7*). Any change in the interfacial surface energy influences the consolidation and void formation during processing.

For discontinuous fiber (i.e. pulp, hammermilled wood) reinforced polymer composites, the interface is vital to achieve mechanical performance because stress is transferred via shear along the fiber surface (*7*). Modification of the interface between the thermoplastic and cellulose fiber governs the performance of the composite material by dictating the critical length (l_t) needed to transfer stress (σ) for a given fiber diameter (d): $l_t/d = (\sigma \, _c E_f/(2\tau_{int} E_c))$ where τ_{int} is the interfacial shear strength (*7*). Two methods have been utilized to increase the interfacial shear strength within composites by influencing the surface chemistry (see detailed reviews (*12-14*)); application of coupling/compatibilizng agents or chemical modification of the fiber surface. While a variety of coupling agents have been introduced, such as isocyanates (*15*) and silanes (*16,17*), the most frequently studied agents involve anhydrides grafted to polymers (*18-22*). Felix

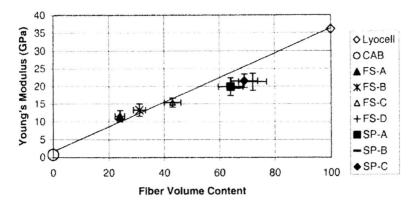

Figure 1. Modulus versus fiber content of lyocell tow-based composite panels produced from film stacking (FS) and solution prepeg (SP). Reproduced with permission from reference (8). Copyright 2000 Springer.

and Gatenholm demonstrated that maleic anhydride graft polypropylene (MAPP) covalently bonded through ester bonds to cellulose fiber surfaces (2). The block-like architecture of the grafted copolymer was found to change the surface energy of cellulose while the polypropylene (PP) chains diffused and entangled into the PP matrix. Lower molecular weight (MW) MAPP did not have the same effect as high MW MAPP for peel strength values of cellulose-PP systems, but all treated surfaces displayed increased strength values over the untreated cellulose surface (23). Microscopic evidence has supported the hypothesis (based on chemical analysis results) that the maleated coupling agent is located at the interface between fiber and matrix (19). However some lubricants were shown to cleave the coupling agent from the fiber surface (24).

Chemical modification of wood via acetylation was first reported by Stamm and Tarkow (25,26) as a method to achieve dimensional stability. Much work has continued in this area (27), showing that esterification of cellulose fiber surfaces influences the surface energies (28) and the adhesion, which results in increased mechanical properties of composites (29). Moreover, if mild reaction conditions are applied during esterification, the surface of the fiber becomes modified relative to the bulk of the fiber as reported by Baiardo et al. for steam-exploded flax fiber (30). Although acetylation of lyocell surfaces did not achieve an increase in mechanical performance for continuous fiber reinforced CAB composites, a favorable response was noted for discontinuous acetylated cellulose fiber (31). Steam-exploded wood that was acetylated increased the mechanical properties of the CAB composites, while unmodified steam-exploded wood caused a decline in strength and reduced the rate of modulus gain (Figure 2). Also evident is the loss in ductility (elongation at break) for the fiber-filled composites with or without interfacial modification (Figure 2).

82

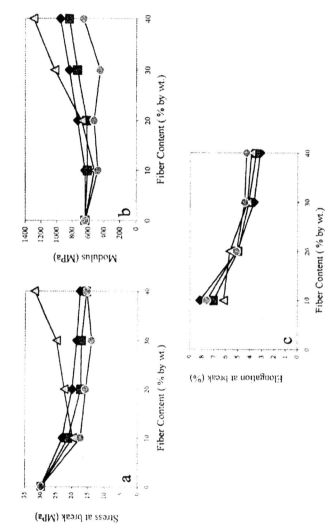

Figure 2. Effect of fiber content on (a) stress, (b) modulus, and (c) elongation at break of CAB filled with various cellulose fibers. ◇- water extracted fiber, ■- alkali extracted fiber, ▲-acetylated and alkali extracted fiber, o- commercial oat a filler[a]. Reproduced with permission from reference (31). Copyright 1999 John Wiley & Sons, Inc.

Typically, the matrix material has high ductility relative to the fiber. Inclusion of the fiber decreases the ductility by serving as a nucleating point for crack propagation.

There is still a possibility to have increases in strength, modulus, and ductility by reducing the size of the reinforcing agent to the nanoscale. Nanoscale particles have dimensions that are smaller than the critical defect size that would act as stress concentrators. Additionally, for many nanocomposites ductility can increase more than that of the unfilled matrix. Pathways related to this increase in ductility are related to mechanisms in rubber toughened epoxies-an increase in volume of material undergoing deformation, as well as strain softening and hardening from shear yielding and crazing (32).

Cellulosic nanocomposites have been created with the addition of isolated nanowhiskers to aqueous latex dispersions or melt extruded composites. It was shown that latex casting is a superior processing method because it allows for hydrogen bonding to develop among the cellulose nano-crystals and matrix material (33). The effect of the nanowhiskers is most evident at temperatures above the glass transition in the rubbery plateau-region of the matrix material. Nanowhiskers can improve the modulus by two orders of magnitude for latex-based composites for this region (34). Additionally this effect is achieved at low loading levels. The reinforcing mechanism has been characterized as a percolation effect where the nanowhiskers form a network from one surface of the composite to the other (35, 36). While there is no physical overlap of the nanowhiskers, the matrix material no longer behaves as the neat system. High specific surface area of the nanoparticles creates a continuous interfacial zone at low levels of nanoparticle loading (32).

The nanowhisker isolation process occurs by acid hydrolysis of the most accessible regions (i.e. the amorphous component) (Figure 3, left schematic). Acid hydrolysis is stopped before complete (glucosylation/depolymerization) of the non-accessible cellulose (i.e. crystalline component). As reported in other chapters in this book, nanowhiskers are purified and recovered by dialysis and freeze-dried before being compounded with matrix components. Matsumura et al. demonstrated an alternative approach to nanocomposite formation that parallels the formation of composites from nanowhiskers and matrix but avoids the acid hydrolysis, whisker-recovery, and compounding steps (37, 38). Cellulose nanocomposites were produced from the heterogeneous esterification of pulp fiber in a non-swelling medium. Esterification (like acid hydrolysis) occurred at the most accessible regions of the pulp fiber, while maintaining unmodified nanodomains of the crystalline cellulose component (Figure 3, right schematic). Based upon atomic force microscopy studies using an enzymatic etching technique, the crystalline nanodomains were found to be evenly distributed within the thermoplastic cellulose ester (38). These composites displayed properties that were typical of other cellulose nano-composites: optical clarity with increases in ductility and mechanical properties (37). In this case, no interfacial modifications were necessary because the cellulose ester was chemically linked to the cellulose nanowhiskers (Figure 3, right schematic).

Hence, by controlling processing conditions (the degree of solvent induced swelling) Matsumura et al. demonstrated that a strong interface can be created without the isolation and subsequent modification of cellulose nanowhiskers (*37,38*).

Reactive Processing of Lignocellulosics

During compounding, extrusion, or injection molding of fiber and thermoplastic the interface may be modified by the introduction of reactants that are activated at elevated processing temperatures. The processing equipment simultaneously acts as reactor where enough thermal energy is generated to initiate reactions of additives, such as peroxides (*39-41*) that can graft polymers onto the fiber surface, initiate the polymerization of monomers, or compatibilize the fiber surface. Thermal energy during processing may also cause incipient degradation of fiber, matrix, or additive. The former follows known chemical pathways for modification, while the latter creates additional mechanisms for interfacial modification.

The polymerization of monomers on the fiber surface during processing is often difficult with much of the monomer forming homopolymer. Wirjosentono et al. provided two schemes of (a) homopolymerization and (b) graft polymerization with their investigation of polymerization of acrylic acid during extrusion of oil palm fiber filled polypropylene (*42*). In contrast, Dorgan and Braun modified cellulose surfaces with initiators that would polymerize lactide during extrusion onto the cellulose surface (*43*). Maximized surface modification occurred when low molecular weight prepolymers were incorporated into the extruder.

Another route to add functionality to the polyolefin is through oxidative reactions ranging from ozone, plasma, open flame, irradiative and thermal treatments, which can be added into compounding and extrusion. Hedenberg and Gatenholm found that polyethylene could be chemical bonded to cellulose fiber surfaces after exposure of the polyethylene to ozone gas through a proposed mechanism of peroxide intermediates (*44*). Renneckar et al. investigated interfacial modification of lignocellulosic/polyolefin composites through reactive steam-explosion co-processing (*45-49*). Hardwood chips were steam-exploded in the presence of polyolefins with steam temperatures reaching 230°C. [At this high temperature, in the presence of moisture, there is (a) acid catalyzed depolymerization of polysaccharides, lignin, and lignin carbohydrate linkages; (b) incipient oxidation of polyolefin; and (c) mobility of the biopolymers within wood.] Co-processed wood and polyolefin were of uniform fibrous form (dependent on the melt rheology of thermoplastic (*46*)) with the surface of the fiber modified by the polyolefin. A change in surface character was illustrated by dipping the modified wood fiber bundle into water (Figure 4) (*47*). Water formed a contact angle greater than 90° indicated by the depletion of the water away from the fiber bundle surface (Figure 4b).

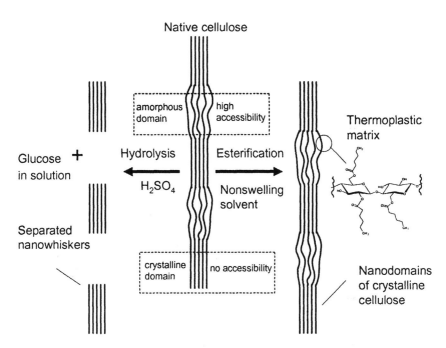

Figure 3. Accessible cellulose component may be hydrolyzed with mineral acid or derivatized in nonswelling solvents, while maintaining unmodified nanodomains of crystalline cellulose.

86

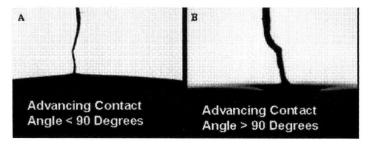

Figure 4. Wetting and non-wetting of fiber observed through the advancing contact angle. A) steam-exploded red oak fiber and B) co-steam-exploded red oak and polypropylene. Reproduced with permission from reference 47. Copyright 2006 Society of Wood Science and Technology.

Co-processed wood and polypropylene (50:50 w/w) were subsequently wetlaid with 10% chopped lyocell fiber and compression molded into composite panels. The composite had an ultimate tensile strength and modulus of 26.7 MPa and 3.4 GPa, respectively, while flexural strength and modulus were 55 MPa and 3.4 GPa. These values are within the range of values reported for wood flour-filled polypropylene composites without compatibilizer (50). However, the polypropylene used for co-steam-explosion processing was low molecular weight ($\sim M_n$: 10,000); mechanical properties were indeterminate for the unfilled polypropylene because the control specimens would break upon loading into the tester. This indicates a substantial gain in strength by coprocessing using steam-explosion. For comparison, an ultimate tensile strength of 27.4 MPa and 2.9 GPa was found for higher molecular weight polypropylene reinforced with lyocell/steam-exploded wood (not reactively co-processed) (51).

Self-Assembly for Interface and Nanocomposite Formation

High performance composite materials have been created using a bottom-up design strategy where the placement of nanoparticles and macromolecules are controlled through self-assembly. With the bottom-up design a wider range of macroscopic properties of the composite material can be derived from similar building blocks by controlling nanoscale structures and interactions. Natural materials already incorporate this design process (52); plant-based materials contain cellulose, lignin and hetero-polysaccharides, but the placement and hierarchical ordering varies, which allows for a diversity of performance. In order to understand nature's template, the self-assembly of model lignin-carbohydrate compounds onto solid cellulose surfaces were studied to design the interface for cellulose reinforced thermoplastic composites from the bottom-up (*53-55*). In these studies, both modified xylans and pullulans were investigated as to their self-assembling properties onto a well-defined model surface representing cellulose (Langmuir-Blodgett films of cellulose regenerated from the silylated derivative). Increasing the substitution with hydrophobic moieties was found to increase the surface activity of the polysaccharidic copolymers as noted by the rapid decline of the critical aggregation concentration (CAC) with concentration (*54*). Further, the docking of the polysaccharidic copolymers from aqueous solutions onto cellulose surfaces was monitored using surface plasmon resonance. Derivatizing the pullulan with bulky hydrophobic substituents (abietic acid) resulted in the deposition of more amphiphilic mass per square nanometer on the cellulose surface (*55*). From these studies it was concluded (a) that the surface of cellulose can be irreversibly modified by self-assembled amphiphilic polymers; and (b) the interfacial thickness is controllable at the nanolevel with low substitution levels (DS of 0.027).

Another strategy involving self-assembly to modify cellulose surfaces is through the adsorption of ionic polymers (*56*). Cellulose has an anionic functionality in aqueous solutions and as a result cationic polymers are attracted to the surface. The configuration of the adsorbed cationic polymer is different than that of the amphiphilic polymers. Strong ionic attractions pull the polyelectrolyte from the aqueous solution and the polymer can dock on the surface with a flat conformation (Figure 5). In contrast, surface activity arising from the non-polar substituents pushes amphiphilic polymers to the surface and the polymers tend to adsorb with loop and tail structures away from the surface (Figure 5).

The adsorption of cationic polymers onto a cellulose surface is illustrated with cationic starch adsorbing to a regenerated cellulose fiber (lyocell) (Figure 6). Atomic force microscopy (AFM) in tapping mode revealed that the cellulose structure aligned parallel to the fiber axis of lyocell is no longer evident after cationic starch (0.5% w/w aqueous solution, Emsland-Starke Gmbh) adsorbs to the fiber surface. The cellulose surface can be considered irreversibly modified with cationic starch; the cationic starch is not removed after washing the fiber with deionized water. Hence, cationic polymers in aqueous solutions adsorb to the cellulose surface by attractive forces and change the surface character.

Adsorbed polyelectrolytes can reverse the existing surface charge of the substrate. This process can be exploited to create nanocomposites with controlled placement of macromolecular structures using the layer-by-layer (LBL) sequential adsorption technique. This multilayer process uses sequential adsorption of oppositely charged polyelectrolytes to create composite materials containing polymers or nanoparticles (*57*). Specifically, a polycation adsorbs to a substrate creating a positively charged surface, then rinsed removing loosely adsorbed material, and the substrate is then placed into a polyanion solution, which adsorbs to the surface reversing the charge to a negatively charged

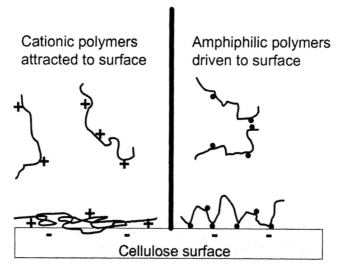

Figure 5. Idealized comparison of the docking process for cationic polymers and amphiphilic polymers onto cellulose surfaces in aqueous solutions. Cationic polymers are attracted (pulled) to cellulose surfaces via ionic and ion- dipole interactions with relatively flat conformations (salt concentration critical). Amphiphilic polymers are surface active towards cellulose surfaces creating loop and tail structures (scheme shows adsorption below CAC).

surface. The process is repeated multiple times developing a multilayer material that is self-assembled layer by layer.

Cellulose nanowhiskers are well suited for the LBL-assembly method because they have a negatively charged surface from sulfate ester groups as a result of isolation with sulfuric acid. Cellulose nano-whiskers isolated from dissolving pulp with a method adapted from Edgar and Gray (58) were used in the LBL process with cationic starch. A cleaned glass cover slip was used as the substrate with cationic starch (0.5% w/w water) and cellulose nanowhiskers (0.5% w/w water) to form a single bi-layer. The glass cover slip was placed in a sealed container and exposed to the cationic starch solution for 5 minutes via flow from a syringe pump. After the 5 minutes Milli-Q water (dispensed via the syringe pump) was used to rinse the glass cover slip. Next, cellulose nanowhiskers were passed over the cover slip for 5 minutes. Finally, Milli-Q water was used to rinse the glass cover slip. Cellulose nanowhiskers clearly adsorbed to the glass cover slip that was pretreated with the oppositely charged starch (Figure 7). Overlapping cellulose nano-whiskers are evident with a change in maximum height of the z-axis of 25 nanometers. While mechanical properties of the film have yet to be studied, hydrogen bonding among the cellulose nanowhiskers, as well as ionic cross-links with the starch serve to create a uniform reinforcement on the nanoscale. In other words, the interface between oppositely charged polyelectrolytes is strong due to the nature of the Coulombic interactions.

A cellulose nanowhisker and cationic starch film with micron dimensions was created from 50 bilayers. Layering occurred by transferring a glass slide by hand to the solutions (0.5% cationic starch and 0.5% cellulose nanowhiskers) in sequential manner. Once 50 bilayers were obtained, the film was carefully peeled off the slide in water and floated onto teflon. The film was then air dried, sputter coated with 8nm of gold-palladium and imaged in a field emission scanning electron microscope with an accelerating voltage of 1kV. The layered structure (highlighted by arrows) is evident at the broken surface (Figure 8). An edge view of a 50 bilayer free standing film reveals the thickness to be 5.6 microns (or 112 nm per bilayer) (inset Figure 8). Layer thickness is greater than reported values for charged polyelectrolytes (59) and will be further studied.

Conclusions

The influence of the interface on cellulose composite properties was reported for two types of geometries: continuous lyocell fibers and wood fibers. Continuous fiber composites attained high mechanical performance without interfacial modification. For discontinuous fibers, interface modification was necessary in order for the fiber to serve as a reinforcing agent. However,

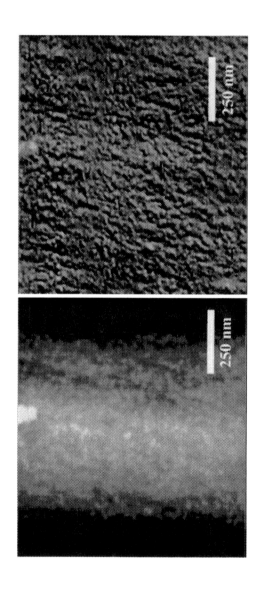

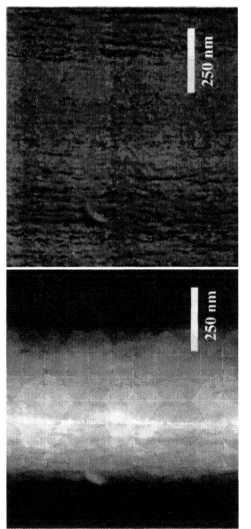

Figure 6. AFM Tapping mode height (left) and phase (right) of (a) unmodified lyocell fiber surface and (b) lyocell fiber surface with adsorbed cationic starch. Note height image in (b) contains spherical clusters, while aligned cellulose structure is masked in phase image. Scan size: 1 μmx1 μm.

Figure 7. AFM Tapping mode height images of a layer of cellulose nanowhiskers adsorbed to cationic starch. (a) Scan size: 10 μm x 10 μm; Z-axis maximum: 30 nm. (b) Scan size: 2 μm x 2μm; Z-axis maximum: 25 nm.

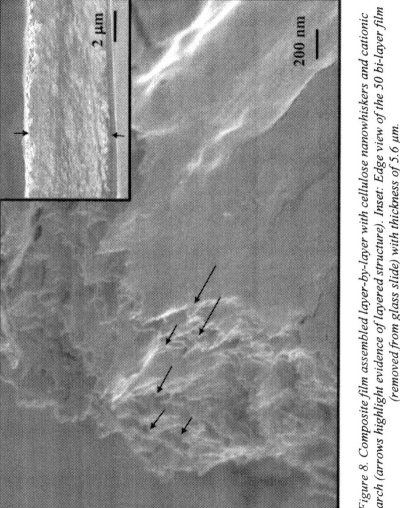

Figure 8. Composite film assembled layer-by-layer with cellulose nanowhiskers and cationic starch (arrows highlight evidence of layered structure). Inset: Edge view of the 50 bi-layer film (removed from glass slide) with thickness of 5.6 µm.

ductility of the composite decreased relative to the neat matrix. By decreasing the size of the reinforcement to the nanoscale both mechanical properties and ductility were improved simultaneously.

Interfacial modification for cellulose composites was achieved by three alternative methods through processing and self-assembly. Modifying cellulose pulp in a non-swelling solvent led to a partially substituted cellulose that contained native cellulose I crystalline domains covalently bonded to the derivatized amorphous component. Additionally, the steam-explosion of wood in the presence of a thermoplastic changed the surface property of the wood fiber, while also modifying the arrangement of the constitutive biopolymers. Furthermore, interfaces were modified by the self-assembly of amphiphilic polymers and polyelectrolytes interacting through secondary forces and electrostatic forces, respectively. Slight derivatization of pullulan with abietic acid (DS 0.027) substantially impacted the adsorption behavior, creating an interfacial region on the cellulose surface by the adsorbed amphiphilic polysaccharides. Ionically-linked interfaces were created with the self-assembly of cationic starch. By using the layer-by-layer approach of reversing surface charge sequentially, cellulose composites of anionic cellulose nanowhiskers and cationic starch were created with controlled nanostructure. These approaches to interface modification expand the avenues for strong cellulose composites outside of the use of traditional coupling agents.

References

1. Schut, J.H. *Plastic. Technol.* September 2005, p 1.
2. Felix, J.; Gatenholm, P. *J. Appl. Polym. Sci.* **1991**, *42*, 609-620.
3. Sherman L.M. *Plastic. Technol.* July 2004, p 1.
4. Sturcova, A.; Davies, G. R.; Eichhorn, S. J. *Biomacromolecules* **2005**, *6*, 1055-1061.
5. Gordon, J.E.; *The new science of strong materials*; Princeton University Press: Princeton, NJ, 1984; pp 1-288.
6. Glasser, W. G.; Rials, T. G.; Kelley, S. S.; Dave, V. *In Cellulose Derivatives*; Heinze, T.; Glasser, W.G., eds.; ACS Symposium Series 688; American Chemical Society, Washington D.C., 1998; pp 265-282.
7. Batch, G. In *Fundamentals of Interfacial Engineering*; Stokes, R.J.; Evans, D.F., Eds; Advances in Interfacial Engineering Series; Wiley-VCH, New York, NY, 1997; pp 347-398.
8. Seavey, K.; Ghosh, I.; Davis, R.; Glasser, W.G. *Cellulose* **2001**, *8*, 149-159.
9. Seavey, K.; Glasser, W.G. *Cellulose* **2001**, *8*,161-169.
10. Franko, A.; Seavey, K.; Gumaer, J.; Glasser, 'W.G. *Cellulose* **2001**, *8*, 171-179.
11. Bourban, C.; Karamuk, E.; De Fondaumiere, M. J.; Ruffieux, K.; Mayer, J.; Wintermantel, E. *J. Environ. Polym. Degr.* **1997**, *5*, 159-166.

95

12. Gauthier, R.; Joly, C.; Coupas, A.; Gauthier, H.; Escoubes, M. *Polym. Composite.* **1998**, *19*, 287-300.
13. Lu, J.Z.; Wu, Q.; McNabb, H.S. *Wood Fiber Sci.* **2000**, *32*, 88-104.
14. George, J.; Sreekala, M.; Thomas, S. *Polym. Eng. Sci.* **2001**, *41*, 1471-1485.
15. Maldas, D.; Kokta, B.V.; Daneault, C. *J. Appl. Polym. Sci.* **1989**, *37*, 751-775.
16. Matuana, L.M.; Balatinecz, J.J.; Park, C.B.; Sodhi, R.N.S. *Wood Sci. Technol.* **1999**, *33*, 259-270.
17. Bengtsson, M.; Gatenholm, P.; Oksman, K. *Compos. Sci. Technol.* **2005**, *65*, 1468-1479.
18. Oksman, K. *Wood Sci. Technol.* **1996**, *30*, 197-205.
19. Oksman, K.; Lindberg, H.; Holmgren, A.; *J. Appl. Polym. Sci.* **1998**, *69*, 201-209.
20. Balatinecz, J.J.; Sain, M.M. *Macromol. Symp.* **1998**, *135*, 167-173.
21. Qiu, W.; Zhang, F.; Endo, T.; Hirotsu, T. *J. Appl. Polym. Sci.* **2003**, *87*, 337-345.
22. Paunikallio, T.; Kasanen, J.; Suvanto, M.; Pakkanen, T.T. *J. Appl. Polym. Sci.* **2003**, *87*, 1895-1900.
23. Felix, J.; Gatenholm, P. *J. Appl. Polym. Sci.* **1993**, *50*, 699-708.
24. Harper, D.; Wolcott, M. *Compos. Part A Appl. Sci.* **2004**, *35*, 385-394.
25. Stamm, A.J.; Tarkow, H. *J. Phys. Colloid Chem.* **1947**, *51*, 493-505.
26. Stamm, A.J.; Tarkow, H. U.S. Patent 2417995, 1947.
27. Rowell, R. *Holzforschung* **1990**, *44*, 263-269.
28. Garnier, G.; Glasser, W.G. *Polym. Eng. Sci.* **1996**, *36*, 885-894.
29. Mahlberg, R.; Paajanen, L.; Nurmi, A.; Kivisto, A.; Koskela, K.; Rowell, R. *Holz Roh Werkst.* **2001**, *59*, 319-326.
30. Frisoni, G.; Baiardo, M.; Scandola, M.; Lednicka, D.; Cnockaert, M. C.; Mergaert, J.; Swings, J. *Biomacromolecules* **2001**, *2*, 476-482.
31. Glasser, W.G.; Taib, R.; Jain, R.K.; Kander, R. *J. Appl. Polym. Sci.* **1999**, *73*, 1329-1340.
32. Schadler, L. *In Nanocomposite Science and Technology*; Ajayan, P.M.; Schadler, L.;Braun P.V., Eds.; Whiley-VCH: Weinheim, 2003, pp 77-153.
33. Brechet, Y.; Cavaille, J.Y.; Chabert, E.; Chazeau, L.; Dendievel, R.; Flandin, L.; Gauthier, C. *Adv. Eng. Mater.* **2001**, *3*, 571-577.
34. Favier, V.; Canova, G.R.; Cavaille, J.Y.; Chanzy, H.; Dufreshne, A.; Gauthier, C. *Polym. Adv. Technol.* **1995**, *6*, 351-355.
35. Favier, V.; Canova, G.R.; Shrivastava, S.C.; Cavaille, J.Y. *Polym. Eng. Sci.* **1997**, *37*, 1732-1739.
36. Favier, V.; Dendievel, R.; Canova, G.; Cavaille, J.Y.; Gilormini, P. *Acta Mater.* **1997**, *45*, 1557-1565.
37. Matsumura, H.; Sugiyama, J.; Glasser, W.G. *J. Appl. Polym. Sci.* **2000**, *78*, 2242-2253.
38. Matsumura, H.; Glasser, W.G. *J. Appl. Polym. Sci.* **2000**, *78*, 2254-2261.

96

39. Cousin, P.; Bataille, P.; Schreiber, H. P.; Sapieha, S. *J. Appl. Polym. Sci.* **1989**, *37*, 3057-3060.
40. Sapieha, S.; Allard, P.; Zang, Y. H. *J. Appl. Polym. Sci.* **1990**, *41*, 2039-2048.
41. Nogellova, Z.; Kokta, B. V.; Chodak, I. J. Macromol. Sci. Pure **1998**, 35, 1069-1077.
42. Wirjosentono, B.; Guritno, P., Ismail, H. *Intern. J. Polym. Mater.* **2004**, *53*, 295-306.
43. Dorgan, J.; Braun, B. *PMSE Preprints* **2005**, *93*, 954-955.
44. Hedenberg, P., Gatenholm, P. *J. Appl. Polym. Sci.* **1996**, *60*, 2377-2385.
45. Renneckar, S.; Zink-Sharp, A.; Ward, T.C.; Glasser, W.G. *J. Appl. Polym. Sci.* **2004**, *93*, 1484-1492.
46. Renneckar, S.; Zink-Sharp, A.; Glasser, W.G. *IAWA J.* under review.
47. Renneckar, S.; Zink-Sharp, A.; Glasser, W.G. *Wood Fiber Sci.* under revision.
48. Renneckar, S.; Wright, R.S.; Gatenholm, P.; Zink-Sharp, A.; Glasser, W.G. Wood Sci. Tech. under revision.
49. Renneckar, S.; Johnson, R.K.; Zink-Sharp, A.; Sun, N.; Glasser, W.G. *Compos. Interface.* **2005**, *12*, 559-580.
50. Youngquist, J.A. *Wood Handbook Wood as an Engineering Material*; Forest Products Laboratory: Washington D.C., 1999, pp 10.1-10.13.
51. Johnson, R.K.; Zink-Sharp, A.; Renneckar, S.; Glasser, W.G. *Holzforschung* under review.
52. Fratzl, P.; Burgert, I.; Gupta, H.S. *Phys. Chem. Chem. Phys.* **2004**, *6*, 5575-5579.
53. Esker, A.; Becker, U.; Jamin, S.; Beppu, S.; Renneckar, S.; Glasser, W.G. In *Hemicelluloses: Science and Technology*; Gatenholm, P.; Tenkanen, M. Eds.; ACS Symposium Series 864; American Chemical Society: Washington DC, 2004; pp 198-219.
54. Gradwell, S.; Renneckar, S.; Esker, A.; Heinze, T.; Gatenholm, P.; Vaca-Garcia, C.; Glasser, W.G. C. *R. Biologie* **2004**, *327*, 945-953.
55. Gradwell, S.E. M.S. Thesis, Virginia Polytechnic Institute and State University, Blacksburg, VA, 2004.
56. Maximova, N.; Österberg, M., Laine, J.; Stenius, P. *Colloid Surface A* **2004**, *239*, 65-75.
57. Decher, G.; Eckle, M.; Schmitt, J.; Struth, B. *Curr. Opin. Colloid In.* **1998**, *3*, 32-39.
58. Edgar, C.D.; Gray, D.G. *Cellulose* **2003**, *10*, 299-306.
59. Podsiadlo, P.; Choi, S,; Shim, B.; Lee, J.; Cuddihy, M.; Kotov, N. A. *Biomacrmolecules* **2005**, *6*, 2914-2918.

Nanocomposites Processing
and Properties

Chapter 8

Cellulose Nanocrystals for Thermoplastic Reinforcement: Effect of Filler Surface Chemistry on Composite Properties

Maren Roman[1] and William T. Winter[2]

[1]Department of Wood Science and Forest Products, Virginia Polytechnic Institute and State University, Blacksburg , VA 24061
[2]Cellulose Research Institute and Department of Chemistry, SUNY College of Environmental Science and Forestry, Syracuse, NY 13210

Many properties of a composite material depend on the nature of the interface between the different components. To study the effect of filler surface chemistry, composites of cellulose acetate butyrate and both native and surface-trimethylsilylated cellulose nanocrystals from bacterial cellulose were investigated by differential scanning calorimetry and dynamic mechanical analysis. Both fillers affected the crystallization of the matrix during solution casting, caused a decrease in heat capacity in the composites, an increase in storage and loss modulus, and a decrease in damping. The increase in composite stiffness was larger for the native fillers. Yet, increases in matrix melting temperatures and recrystallization upon heating as well as a larger decrease in damping indicated better filler–matrix compatibility for the silylated crystals.

Introduction

Cellulose nanocrystals are known to generate significant reinforcement in polymeric materials (*1-3*). The reinforcing effect is thought to be due to a percolating network of filler particles stabilized through hydrogen bonds between the filler particles. The same hydrophilic surface chemistry that allows for the filler–filler hydrogen bonding, however, is incompatible with hydrophobic composite matrices.

In an effort to overcome this limitation, we studied the heterogeneous trimethylsilylation of bacterial cellulose nanocrystals for the preparation of crystals with increasingly hydrophobic surfaces (*4*). The current study is concerned with the characterization of polymer nanocomposites of native and surface trimethylsilylated cellulose nanocrystals. The overall objective of the study is to test the hypothesis that surface modification of the crystals provides a route to enhanced filler–matrix interaction in the composites. A butyryl rich CAB was used as hydrophobic matrix. The CAB–cellulose nanocomposites were characterized by differential scanning calorimetry (DSC) and dynamic mechanical analysis (DMA).

Experimental

Bacterial cellulose (Primacel) was provided by the Nutrasweet Kelco Company (now CP Kelco, Chicago, IL). CAB was kindly donated by Eastman Chemical Company, Kingsport, TN. The sample (CAB-500-5) had a number average molecular weight of 59,000 Dalton, a density of 1.18 $g \cdot cm^{-3}$, and a falling ball viscosity of 5 s. The acetyl content of the triester was ~4 wt % and the butyryl content was ~51 wt % with less than one hydroxyl group per four anhydroglucose units. Thus, the polymer was largely cellulose butyrate. The glass transition temperature was reported to be 96 °C. The polymer was in powder form and used as received. 1,1,1,3,3,3-Hexamethyldisilazane (98%) and formamide (99.5%) were purchased from ACROS Organics, anhydrous diethyl-ether from Merck, and acetone from J. T. Baker. All chemicals were reagent grade or higher in purity and used as received, except the acetone, which was dried over 3-Å molecular sieve from ACROS Organics.

Nanocrystal Preparation and Surface Silylation

Cellulose nanocrystals were prepared by hydrolysis with boiling 2.5 N sulfuric acid for two hours as described previously (*5*). The lateral dimensions of the crystals, measured from transmission electron micrographs (*5*), were consistent with previous reports (6) of 30–50 nm and 6–10 nm in width and thickness, respectively. Lengths seen in the transmission electron micrographs ranged from 200 nm to several micrometers.

Trimethylsilylation of the crystal surface was performed heterogeneously with hexamethyldisilazane (HMDSZ) in formamide (4, 7). Freeze-dried cellulose nanocrystals (0.5 g, water content 3%) were readily dispersed in 35 mL of formamide and heated under argon to 70 °C. A large excess (20 mL) of HMDSZ was added to the suspension. The mixture was heated without stirring for 65 min. The silylated crystals were filtered hot, washed with acetone, dried in a vacuum oven at 80 °C, slurried with diethylether in a mortar, and air dried to a free flowing powder. The integrity of the crystals was confirmed by X-ray diffraction using a Rigaku DMAX 1000 system operating on a 12 KW rotating anode generator and nickel filtered CuKα radiation (X = 0.15419 nm). A decrease in crystallite size was not apparent in the X-ray linewidths.

The average degree of substitution (DS) in the silylated crystals was 0.49 as determined by optical emission spectrometry (Perkin Elmer Optima 3300 DV). With the estimate that 18% of the cellulose chains in the crystals are on the surface (4), a DS of 0.49 signifies an oversilylation. The silylation was continued past the point of complete surface derivatization, to ensure that there are no surface hydroxyl groups available for hydrogen bond formation. To ensure that no degradation of the crystals occurred in the temperature range of the experiments, the thermostabilities of both the native and the silylated crystals were measured by thermogravimetric analysis in air. A 2% weight loss due to thermal degradation was recorded at 240 and 243 °C, and the maximum weight loss rates were observed at 288 and 287 °C, respectively.

Nanocomposite Preparation

Fine dispersions of the crystals, at various concentrations, in acetone were prepared by sonication. Corresponding amounts of CAB were added to the dispersions to give a mixture of 20% composite 80% acetone and the mixtures were stirred for several hours. Then the mixtures were poured onto glass plates. Films of about 6 mm thickness, as cast, were prepared using a Teflon doctor blade, and then covered so that the solvent evaporated slowly over a period of several days. The resulting films were transparent and very brittle. To separate the films from the glass plates without breaking, the films were softened by placing the glass plates for 1 h over acetone in a closed desiccator. The plates were then carefully immersed into a water bath. The films separated from the glass plates within 5 min and were dried in air. To remove residual solvent, the films were dried further in a desiccator under vacuum.

Differential Scanning Calorimetry

DSC measurements were performed with a TA Instruments DSC 2920 calorimeter equipped with a liquid nitrogen cooling accessory. The instrument had been calibrated for temperature with mercury, water, and indium at a heating rate of 10 °C·min^{-1} Calibration for energy was done using the area under the melting curve of indium. The purge gas for the sample cell was nitrogen. The sample size was around 20 mg. Two to three thin (0.2–0.3 mm), flat disks were stacked in an aluminum pan and covered with an inverted lid. To insure good thermal contact the lid was pressed down with a metal cylinder. Each DSC measurement consisted of an initial heating run, a cooling run, and a second heating run between 10 and 215 °C. The heating/cooling rate was 10 °C·min^{-1}.

Melting temperatures were measured at the peak maximum and glass transition temperatures were measured at the inflection point. Crystallinities in the composite matrix were estimated using the equation (8):

$$X_c = \frac{1}{(1-w_F)} \frac{\Delta H_f}{\Delta H_f^\circ} \cdot 100$$

where w_F is the weight fraction of the filler in the composite and ΔH_f° is the heat of fusion of the matrix polymer at 100% crystallinity. As an approximation, ΔH_f° was taken as 34 J·g^{-1}, which is the heat of fusion of 100% crystalline cellulose tributyrate (CTB) (9, 10).

Dynamic Mechanical Analysis

Dynamic mechanical measurements were carried out with a TA Instruments DMA model 2980 in the multifrequency mode using a film tension clamp. Prior to the tests, the sample strips of 0.3 by 4 by 20 mm were slowly heated in an oven to 175 °C to eliminate matrix crystallinity. The absence of crystallinity was confirmed by differential scanning calorimetry. During the DMA measurements, samples were heated from 30 to 96 °C in 3-degree increments and from 96 °C to the temperature of sample yielding in 1-degree increments. Each step began with an equilibration time of either 1 min (3-degree steps) or 20 s (1-degree steps) followed by a frequency sweep. The frequencies employed were 150, 100, 30, 10, 3, and 1 Hz. For brevity's sake, only the 10 Hz data will be discussed here. The oscillation amplitude was 10 μm. Experiments were run with autostrain option, which adjusts the static force throughout the experiment according to a chosen autostrain value, the sample stiffness, and the oscillation amplitude. The initial static force and the autostrain value were 0.03 N and 110%, respectively.

Optical Clarity of the Nanocomposites

Figure 1 shows dogbone samples of the CAB–cellulose nanocomposites. The samples were between 0.2 and 0.4 mm thick and mostly clear. We noted a slight opaqueness at higher filler contents, which disappeared upon annealing at 175 °C. Thus, we attributed the opaqueness to matrix crystallinity.

The composites with silylated fillers were smooth at the surface and uniform in color. The composites with native fillers had a grainy appearance. We were unable to detect any particles in the composites by optical microscopy and hence assumed fairly good dispersion of the nanoparticles. Polarized-light microscopy revealed large matrix crystallites in the composites with native crystals, whereas no such crystallites existed in the composites with silylated crystals. Thus, the grainy appearance of the composites with native crystals was likely due to these matrix crystallites. The large crystallites could be a result of transcrystallization, which has been reported for polymer composites containing cellulose fibers (11-17). Transcrystallization has also been postulated for composites of poly(hydroxylalkanoate) and cellulose crystals from tunicates (18).

Surprisingly, polarized-light microscopy revealed an increasing degree of orientation with increasing filler content in the composites with silylated crystals. This preferred orientation of the matrix could be a result of the directional casting process and is only observed for the silylated crystals because of increased solvent–filler affinity with respect to the native crystals.

Double-Melting Behavior of the Matrix

Figure 2 shows the first-heating, cooling, and second-heating DSC curves for unfilled CAB. During the first heating, two melting peaks were observed confirming that the film cast from solution was partially crystalline. The tendency of highly substituted cellulose esters to crystallize is well known (19, 20). The sample did not crystallize upon cooling, and the cooling and second-heating curves only showed the glass transition. The glass-transition was not clearly resolved during the first heating but a steady increase in heat capacity below 100 °C indicated a relaxation in amorphous regions.

Multiple melting endotherms have been observed in many polymers. A short review is given by Runt and Harrison (8). The proposed explanations for this phenomenon can be grouped into two categories. Explanations of the first category are based on morphological effects and assume the existence of different types of crystals in the sample based on, for example, different mechanisms of nucleation (homogeneous/heterogeneous), lattice structures, chain folding/extension, or molecular weight fractions. Explanations of the second category are based on kinetic effects including initial or partial melting followed by recrystallization or reorganization followed by final melting.

Morphological and kinetic effects on multiple melting can be distinguished by DSC experiments with different heating rates. In such experiments, we

104

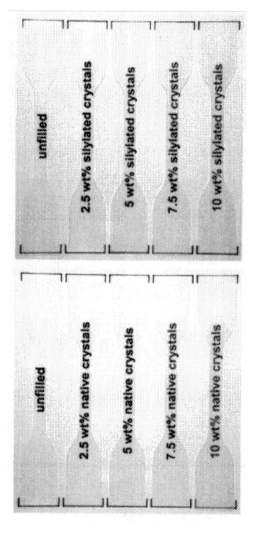

Figure 1. Appearance of CAB–cellulose nanocomposites.

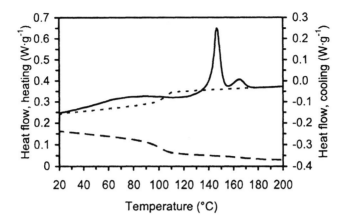

Figure 2. DSC curves for unfilled CAB: first heating (———),
cooling (— —), second heating (- - -).

observed a systematic enlargement of the second melting peak with respect to the first with decreasing heating rate. Furthermore, at very low heating rates (2.5 °C·min⁻¹) an exotherm became discernible between the two melting peaks (4). This behavior is indicative for kinetic effects and, thus, the multiple melting endotherms in our samples are due to a first melting, a recrystallization, and a second melting process of the composite matrix upon heating.

Effect of Nanofillers on Double Melting Behavior

The DSC curves for the composites showed the same characteristics as those for unfilled CAB. The first-heating DSC curves for the composites between 120 and 180 °C are shown in Figure 3. The melting temperatures, heats of fusion normalized to the same matrix mass, and estimated crystallinities are listed in Table I.

Matrix crystallinity decreased slightly with increasing filler content for the native crystals but stayed constant for the silylated ones. In contrast, melting temperatures stayed constant for the native crystals and increased with increasing filler content for the silylated ones. The same was true for the heats of fusion for the second melting. The silylated crystals seemed to facilitate the recrystallization of CAB in the composites upon heating whereas the native crystals appeared to have no effect on CAB recrystallization. In contrast, referring back to the results from polarized-light microscopy, the native crystallites seemed to facilitate crystallization of CAB from solution as manifested in the large matrix crystallites in the composites.

To summarize, both nanofillers affected the crystallization of CAB. The fact that the silylated crystals caused higher melting temperatures in the CAB and an increase in heat of fusion for the second melting process suggests better filler–matrix compatibility for the silylated fillers.

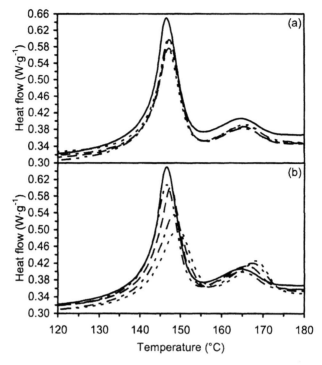

Figure 3. First-heating DSC curves for CAB–cellulose nanocomposites containing (a) native and (b) silylated crystals, filler content in wt %: 0 (——), 2.5 (– - -), 5 (– –), 7.5 ( ), 10 (- - -).

Table I. Melting temperatures, T_m, heats of fusion, ΔH_f, and crystallinities, X_c, from the first-heating DSC curves.

w_F	T_{m1} (°C)	T_{m2} (°C)	ΔH_{f1}^a (J·g^{-1})	ΔH_{f2}^a (J·g^{-1})	ΔH_f^a (J·g^{-1})	X_c (%)
0.000	146.5	164.2	14.8	3.7	18.5	54.5
Native						
0.025	147.1	165.0	14.1	3.4	17.5	51.5
0.050	147.2	164.7	13.5	3.3	16.8	49.3
0.075	147.1	165.5	12.8	3.4	16.2	47.7
0.100	146.9	165.5	12.5	3.3	15.8	46.4
Silylated						
0.025	146.6	164.5	14.0	3.7	17.7	51.9
0.050	147.6	165.9	13.8	4.7	18.5	54.3
0.075	148.2	167.2	13.0	5.2	18.2	53.5
0.100	149.3	168.3	12.3	5.8	18.1	53.1

^{a} normalized to the same matrix mass

Source: Reproduced with permission from *Journal of Polymers and the Environment*, 2002, *10(1/2)*. Copyright 2002 Kluwer Academic.

Effect of Nanofillers on Glass Transition and Heat Capacity

The cooling DSC curves for the composites are shown in Figure 4. The glass transition temperatures, T_g, and the heat capacity increments at the glass transition, ΔC_p, are listed in Table II. The glass transition of the composite matrix seemed to be unaffected by the fillers. The transition occurred in all samples at ~101 °C during cooling and at ~107 °C during second heating. The average change in heat capacity was 0.38 $J{\cdot}K^{-1}{\cdot}g^{-1}$ during cooling and 0.33 $J{\cdot}K^{-1}{\cdot}g^{-1}$ during second heating. The fact that T_g did not change with filler content suggests that neither of the fillers affected the onset of translational and rotational backbone motions in the composite matrix.

The vertical upward shift of the cooling DSC curves (Figure 4) signified a decrease in specific heat capacity with introduction of the fillers. In the case of the native crystals, the decrease was sudden at 2.5 wt % filler loading and only small changes were observed at higher filler loadings. In the case of the silylated crystals the decrease in specific heat capacity was gradual with increasing filler content. A smaller heat capacity in the composites with respect to unfilled CAB was expected because the heat capacity of cellulose crystals is lower than that of CAB. However, the experimental heat capacities of the composites were lower than the weighted averages of the heat capacities of the two components. The discrepancy between the predicted and experimental data indicates that the heat capacity of one or both components changed upon mixing. The heat capacity of the cellulose crystals is presumably unaffected by the presence of the matrix. Therefore the discrepancy must be due to a change in the heat capacity of the matrix in the interphase.

Effect of Nanofillers on Storage and Loss Modulus

The storage modulus, E', and loss modulus, E'', for different filler contents as a function of temperature are shown in Figure 5. Both, storage and loss modulus were higher at higher filler contents indicating good interfacial contact in the composites so that load transfer from matrix to filler can occur. The temperature dependence of E' and E'' did not change significantly with introduction of the fillers. The storage modulus decreased slightly with temperature up to about 100 °C and then dropped by more than two orders of magnitude due to the glass transition of the matrix to values below the instrumental stiffness limits. The drop in E' shifted to higher temperatures with increasing filler content. The loss modulus increased slightly with temperature up to 100 °C and showed a peak before it dropped by one order of magnitude. The peak in E'' increased in magnitude and shifted to higher temperatures with increasing filler content.

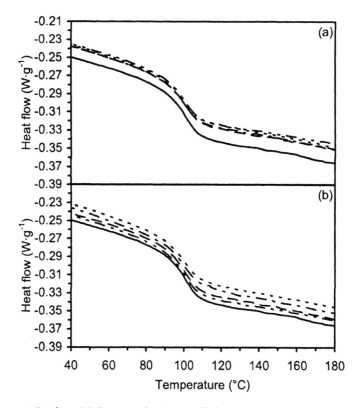

Figure 4. Cooling DSC curves for CAB-cellulose nanocomposites containing (a) native and (b) silylated crystals, filler content in wt %: 0 (——), 2.5 (— - -), 5 (– –), 7.5 (— -), 10 (- - -).

Table II. Glass-transition temperatures, T_g, and heat capacity increments at the glass transition, ΔC_p, from the cooling and second-heating DSC curves.

	Cooling		Second heating	
w_F	(°C)	$\Delta C_p{}^a$ $(J \cdot K^{-1} \cdot g^{-1})$	(°C)	$\Delta C_p{}^a$ $(J \cdot K^{-1} \cdot g^{-1})$
0.000	100.8	0.37	106.7	0.33
Native				
0.025	99.4	0.37	106.2	0.34
0.050	100.4	0.37	106.4	0.36
0.075	100.9	0.38	106.6	0.32
0.100	101.4	0.37	106.8	0.34
Silylated				
0.025	101.5	0.38	106.4	0.34
0.050	100.6	0.38	106.7	0.36
0.075	100.2	0.39	106.5	0.31
0.100	100.9	0.40	107.3	0.31

[a] normalized to the same matrix mass

The values of the storage and loss modulus at the arbitrary temperatures 81 and 119 °C for different filler contents are listed in Table III. At 81 °C, the storage modulus had increased by a factor of 1.9 for the native crystals and by a factor of 1.7 for the silylated ones at 10 wt % filler content. At 119 °C, the storage modulus had increased by a factor of 10 for the native crystals and by a factor of 7 for the silylated ones at 10 wt % filler content. The superior reinforcement for the native fillers might be due to filler–filler interactions through hydrogen bonding as postulated previously (1-3).

Effect of Nanofillers on Tan δ

Tan δ, also called damping, is the ratio of loss to storage modulus. A selection of tan δ curves is shown in Figure 6. With increasing filler content, the tan δ peak, related to the glass transition and α relaxation of CAB, shifted to higher temperatures, broadened, and decreased in magnitude. The tan δ peak temperatures, T_α, and peak heights, tan δ_{max}, are listed in Table III. In DSC experiments, the glass transition temperature stayed constant with increasing filler content (Table II). DSC glass transition temperatures correspond to DMA glass transition temperatures at 0 Hz oscillation. The difference between the glass transition temperatures measured by DSC and the DMA values reported here lies in the frequency dependence of the glass transition at any given filler content.

110

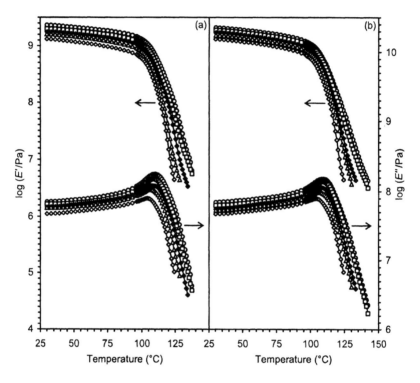

Figure 5. Storage modulus, E', and loss modulus, E", of CAB-cellulose nanocomposites as a function of temperature: (a) native crystals, (b) silylated crystals; filler content in wt %: 0 (◇), 2.5 (△), 5 (◆), 7.5 (□), 10 (○).
Source: Reproduced with permission from *Journal of Polymers and the Environment*, 2002, *10(1/2)*. Copyright 2002 Kluwer Academic.

Table III. Storage modulus, *E'*, and loss modulus, *E"*, at 81 and 119 °C, and tan δ peak temperatures, T_α, and peak heights, tan δ_{max}.

w_F	E'(MPa)		E"(MPa)		$T_\alpha(°C)$	tan δ_{max}
	810°C	*119°C*	*81°C*	*119°C*		
0.000	900	22	60	20	125	1.52
Native						
0.025	1150	43	71	38	126	1.40
0.050	1302	103	77	59	128	1.14
0.075	1505	151	82	80	129	0.86
0.100	1750	223	91	106	131	0.79
Silylated						
0.025	1073	54	68	36	127	1.13
0.050	1197	75	74	47	127.5	1.05
0.075	1364	124	80	70	129	0.79
0.100	1509	159	88	80	131	0.68

Source: Reproduced with permission from *Journal of Polymers and the Environment*, 2002, *10(1/2)*. Copyright 2002 Kluwer Academic.

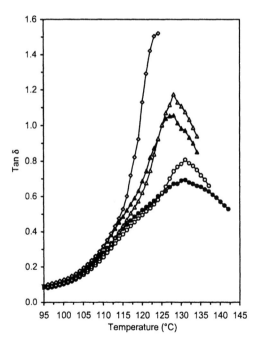

Figure 6. Tan δ curves of CAB-cellulose nanocomposites: native fillers (open symbols), silylated fillers (filled symbols), filler content in wt %: 0 (◇), 5 (△,▲), 10 (○,●). Source: Reproduced with permission from Journal of Polymers and the Environment, *2002,* 10(1/2). *Copyright 2002 Kluwer Academic.*

The decrease in tan δ_{max} was more pronounced for the silylated crystals. However, below the yield temperature for unfilled CAB, the damping with silylated crystals was larger than with native crystals. The sudden increase in damping in the composites with native fillers at the T_α of unfilled CAB could be an indication for weak filler-matrix interactions. If the degree of adhesion is small the matrix will soften at about the same temperature as in the unfilled state and matrix molecules will start to slip along the filler surface. This slippage will result in additional energy dissipation, increasing tan δ. The cross over in the tan δ curves at the yield temperature for unfilled CAB suggests that the silylated filler surface provides stronger interfacial interactions than the more hydrophilic surface of the native crystals.

Conclusions

The objective of the study was to test the hypothesis that surface modification of cellulose nanocrystals provides a route to enhanced filler–matrix interaction in cellulose nanocomposites. The increase in matrix melting temperatures and recrystallization upon heating and the decrease in damping for the silylated crystals suggest better filler–matrix compatibility for the crystals with hydrophobic surface chemistry. The larger reinforcement ability of the native crystals could be a result of hydrogen bonding between the crystals.

Acknowledgements

The project was supported by the USDA Cooperative State Research, Education and Extension Service, grant numbers 96-35501-3454 and 98-35504-6358, and by Eastman Chemical Company.

References

1. Favier, V.; Chanzy, H.; Cavaillé, J. Y. *Macromolecules* **1995**, *28*, 6365-6367.
2. Helbert, W.; Cavaillé, J. Y.; Dufresne, A. *Polym. Compos.* **1996**, *17*, 604-611.
3. Chazeau, L.; Cavaillé, J. Y.; Canova, G.; Dendievel, R.; Boutherin, B. *J. Appl. Polym. Sci.* **1999**, *71*, 1797-1808.
4. Grunert, M. Ph.D. thesis, SUNY College of Environmental Science and Forestry, Syracuse, NY, 2002.
5. Roman, M.; Winter, W. T. *Biomacromolecules* **2004**, *5*, 1671-1677.
6. Tokoh, C.; Takabe, K.; Fujita, M.; Saiki, H. *Cellulose* **1998**, *5*, 249-261.
7. Harmon, R. E.; De, K. K.; Gupta, S. K. *Carbohydr. Res.* **1973**, *31*, 407-409.

8. Runt, J.; Harrison, I. R. In *Polymers, Part B. Crystal Structure and Morphology;* Fava, R. A., Ed.; Methods of Experimental Physics; Academic Press: New York, NY, 1980; Vol. 16, pp 287-337.

9. Mandelkern, L.; Flory, P. J. *J. Am. Chem. Soc.* **1951**, *73*, 3206-3212.

10. Piana, U.; Pizzoli, M.; Buchanan, C. M. *Polymer* **1995**, *36*, 373-380.

11. Gray, D. G. *J. Polym. Sci., Polym. Lett. Ed.* **1974**, *12*, 509-515.

12. Quillin, D. T.; Caufield, D. F.; Koutsky, J. A. *J. Appl. Polym. Sci.* **1993**, *50*, 1187-1194.

13. Felix, J. M.; Gatenholm, P. *J. Mater. Sci.* **1994**, *29*, 3043-3049.

14. Amash, A.; Zugenmaier, P. *Polym. Bull. (Berlin)* **1998**, *40*, 251-258.

15. Amash, A.; Zugenmaier, P. *Polymer* **1999**, *41* 1589-1596.

16. Son, S.-J.; Lee, Y.-M.; Im, S.-S. *J. Mater. Sci.* **2000**, *35*, 5767-5778.

17. Zafeiropoulos, N. E.; Baillie, C. A.; Matthews, F. L. *Compos. A, Appl. Sci. Manuf.* **2001**, *32A*, 525-543.

18. Dufresne, A.; Kellerhals, M. B.; Witholt, B. *Macromolecules* **1999**, *32*, 7396-7401.

19. Boy, R. E., Jr.; Schulken, R. M., Jr.; Tamblyn, J. W. *J. Appl. Polym. Sci.* **1967**, *11*, 2453-2465.

20. Kozlov, P. V.; Rodionova, M. I. *Polym. Sci. U.S.S.R.* **1966**, *8*, 1083-1087.

Chapter 9

The Structure and Mechanical Properties of Cellulose Nanocomposites Prepared by Twin Screw Extrusion

Aji P. Mathew[1], Ayan Chakraborty[2], Kristiina Oksman[1], and Mohini Sain[3]

[1]Department of Engineering Design and Materials, Norwegian University of Science and Technology, Trondheim, Norway
[2]Department of Chemical Engineering and Applied Chemistry, University of Toronto, Toronto, Ontario, Canada
[3]Centre for Biocomposites and Biomaterials Processing, Faculty of Forestry and Chemical Engineering, University of Toronto, Toronto, Ontario, Canada

The goal of this work has been to prepare cellulose nanocomposites of polylactic acid (PLA), cellulose nano whiskers (CNW) and microfibers (MF). Nanocomposites were prepared by pumping an aqueous dispersion of MF and CNW into the PLA during extrusion. The prepared materials were studied using different microscopy methods (TEM, AFM, SEM), X-ray, dynamic mechanic thermal analysis (DMTA) and conventional mechanical testing. The MF was shown to form a network of fibrils while CNW existed as needle shaped crystallites after the isolation process. DMTA and tensile tests indicated no significant improvement in mechanical properties of the composites. This may be attributed to poor dispersion of microfibres and nanowhiskers in PLA.

Introduction

Cellulose is the most abundant, renewable and biodegradable natural polymer on earth. Cellulose fibers are present in plant cell walls in combination with hemicelluloses, lignin, waxes etc (*1*). These fibers consists of bundles of microfibrils where the cellulose chains are stabilized laterally by hydrogen bond between hydroxyl groups. Cellulose microfibrils can be separated from various sources by chemical and mechanical treatments. This cellulose fibril can be about 5-10 nm in diameter and the length varies from 100 nm to several micrometers depending on the source (*1-3*). Each microfibril consists of monocrystalline cellulose domains linked by amorphous domains. On acid hydrolysis the microfibrils undergo transverse cleavage along the amorphous regions into microcrystalline cellulose or whiskers. Due to the near perfect crystalline arrangement of whiskers they have high modulus and act as efficient reinforcing materials (*4*). Properties of cellulose crystallites from earlier reports are shown in Table I (*2,3*). However, the high reinforcing potential of the microfibrils and the whiskers has not been fully exploited and utilized in a commercial scale yet, even though attempts are being made in this direction.

Table I. Properties of Cellulose Whiskers

Property	*Cellulose crystallites*
Length (nm)	300-600
Diameter (nm)	5-10
Aspect Ratio (l/d)	20-60
Tensile Strength (MPa)*	10000
E-Modulus (GPa)*	150-250

*ref. (*2,3*)

Cellulosic reinforcements when combined with polymers from renewable resources are the potential and effective way to produce the so-called green materials with improved performance. Biopolymers including cellulosic plastics obtained from wood (CA, CAB), starch, polylactic acid (PLA) derived from corn and poly hydroxyl alkanoates (PHAs) produced by bacteria are very interesting matrix polymers in this context. (*5-9*). There are several works on cellulose based nanocomposites and have reported exceptional properties. Dufresne and coworkers have prepared nanocomposites by solvent casting of various water soluble polymers using wheat straw, tunicin, chitin and sugar beet as reinforcements (*10-14*). They have reported the tangling effect of microfibrils from sugar beet on poly (styrene-co-butyl acrylate) (*15*). Nakagaito et al. have separated microfibrils from kraft pulp by a homogenizing process and used it as reinforcement in PF resin (*16, 17*). Cellulose nanocomposites based on cellulose nanocrystals from bacterial cellulose and cellulose acetate butyrate prepared by solution casting was reported by Grunert and Winter (*18*). Wu et al. have reported about elastomeric PU/cellulose nanocomposites prepared by in-situ polymerization (*19*). However, all these studies on cellulose nanocomposite processing have been limited to laboratory scale, and focused on solvent casting.

In this work, melt extrusion using a twin screw extruder was explored as a technique of preparing cellulose nanocomposites. The advantage of melt extrusion lies in the fact that unlike solvent casting, this method can be scaled up to an industrial level. For this purpose, two different cellulose reinforcements, cellulose nanowhiskers (CNW) and microfibres (MF), were considered in a polylactic acid (PLA) matrix. PLA is a versatile polymer made from renewable agricultural raw materials and is fully biodegradable. Furthermore, PLA possesses excellent processibility, good stiffness and strength (*20, 21*). The main drawbacks with PLA are low toughness and thermal stability. The reinforcements used in this study are based on wood products, viz, wood pulp and microcrystalline cellulose (MCC), which are commercially available in bulk. We have recently reported the preparation of polylactic acid based cellulose nanocomposites by melt extrusion. In this study DMAc/LiCl solution was used as the dispersion medium for cellulose whiskers. However, this system showed degradation during high temperature processing (*22*). CNW was isolated from MCC by acid hydrolysis and dispersed in aqueous medium. Cellulose whiskers are stiff, individual, crystalline needle shaped entities. The microfibers were separated from wood pulp by cryocrushing and filtration. Microfiber (MF) in this context refers to fibers of cellulose which are 1 tm and less in diameter. The structures of the nanoreinforcements and the nanocomposites were studied using microscopic methods and X-ray diffraction. The material performance was evaluated by dynamic mechanical thermal analysis (DMTA) and conventional tensile testing.

Experimental

Materials

Matrix: Poly Lactic Acid (PLA), Nature Works$_1^{TM}$ 4031 D, was supplied by Cargill Dow LLC, Minneapolis, USA. The density, glass transition temperature (T_g) and melting point are 1.25 g/cm^3, 58 °C and 160 °C respectively. PLA has a molecular weight (Mw) ranging between 195,000 and 205,000 g/mol, and a melt flow index (MFI) of 2-5 g/min.

Reinforcements. Micro Crystalline Cellulose (MCC), VIVAPUR® 105, supplied by J. Rettenmaier & SÖHNE GMBH + CO (Rosenberg, Germany), is commercially available and was used as raw material for the isolation of the whiskers. It is > 93 % pure microcrystalline cellulose and the particle size is between 10-15 µm. Bleached softwood kraft pulp for generating the MFs was supplied by Kimberly-Clark Forest Products Inc., (Terrace Bay, Ontario, Canada). It has the following composition, cellulose 88 %, hemicellulose 11%, Lignin<1 %.

Chemicals. Sulphuric acid (98 %) was used for the acid hydrolysis of MCC. Polyethylene glycol (PEG 1500) was used as a processing aid to decrease the viscosity of the system. These chemicals were purchased from Merck (Schuchardt, Germany).

Preparation of Cellulose Whiskers

Cellulose whiskers were prepared using acid hydrolysis. MCC was dispersed in water (5 wt%), treated with sulphuric acid (65 wt%) and heated at 60 °C during 20 min to achieve complete hydrolysis. The acid was removed by repeated centrifuging and washing with distilled water, followed by dialysis against distilled water, until neutral. The suspension was then collected and sonified in small volumes for 15 min to isolate whiskers. The final aqueous suspension of whiskers was passed through a No. 2 filter to remove any coarse particles before extrusion.

Preparation of Microfibers

Microfibers of cellulose were prepared from bleached northern black spruce kraft pulp by a combination of refining and cryocrushing under liquid nitrogen. The resulting fibers were filtered through a screen of mesh size 60, yielding fibers less than 1 µm in diameter in the filtrate. Details of the microfiber generation process are outlined by Chakraborty et al. (*23*).

Processing of Nanocomposites

The composite materials were compounded using a co-rotating twin screw extruder (Coperion Werner & Pfleiderer ZSK 25 WLE, Germany) with a gravimetric feeding system for dry materials (K-Tron AG, Switzerland) and a peristaltic pump (Drive PD 5006 Heildoph Instruments GmbH, Germany) for dispersed whiskers and microfibers. The pumping speed was calibrated for the used dispersions to ensure accurate feeding of the dispersions. The extrusion was carried out in the temperature range of 165-185 °C and the screw speed was held constant at 100 rpm. The total throughput was 4 kg/hr in all the cases which is very low for an extruder with a total capacity if 50 kg/h. Figure 1 shows a schematic picture of the compounding process. The PLA polymer was fed in zone 1 and the dispersed whiskers and microfibers were pumped into the melt polymer at zone 4. The concentration of whiskers and microfibers in the dispersions was 4 wt% (by weight). The pumping speed was 5 kg/h, which correspond to 200 g/h whiskers / microfibers and resulted in 5 wt% of CNW/MF in the final composition. The processing aid (PEG) was dissolved both in the dispersions and water at a concentration of 4 wt %. The pumping speed when the processing aid was used was 5.2 kg/h. The introduction of aqueous dispersions into the hot polymer melt was difficult, mainly because of the large amount of steam generated. This problem was solved by lowering the screw speed as well as throughput and by increasing the number ventings during the processing. The liquid phase was removed by atmospheric ventings in zones 7 and 8, and by vacuum venting in zone 10. Table II summarizes the processing parameters.

Table II. Processing Parameters

Parameter	Value
Main feeder (kg/hr)	3.8 and 3.6
Pumping rate (kg/hr)	5.0 and 5.2
Screw rotation speed (rpm)	100
Motor load (torque) (%)	55-60
Pressure at the die (bar)	2

The formulations of the different materials during the compounding process and the final composition are summarized in Table III.

Characterization

Microscopy

The morphology of cellulose whiskers and microfibers was studied using a Philips CM30 transmission electron microscope (TEM) and a Nanoscope IIIa

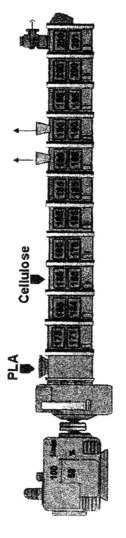

Figure 1. Schematic picture of the compounding process: Feeding of PLA in zone 1, pumping of dispersed of CNW in zone 4 and removal of the water using atmospheric and vacuum ventings in zones 7, 8 and 10.

**Table III. Formulations of Prepared Materials During Processing and
the Final Composition (% by weight)**

Final composition	PLA	PEG	CNW	MF
PLA (100)	100	-	-	-
PLA-CNW (95/5)	95	-	* 120/5	-
PLA-MF (95/5)	95	-		*120/5
PLA-PEG (95/5)	95	5		
PLA-PEG-CNW (90/5/5)	90	5	* 115/5	
PLA-PEG-MF (90/5/5)	90	5		* 115/5

*pumped water which is removed during extrusion.

multimode atomic force microscope (AFM) from Veeco. TEM study was performed at an acceleration voltage of 100 kV. To examine the extent of MCC separation, one droplet of 0.01 % diluted suspension was put on a Cu-grid coated with a thin carbon film. The nano sized whiskers were positively stained in a 1 wt % solution of uranyl acetate (a heavy metal salt) in de-ionized water for 1 min to enhance contrast in TEM. A droplet of the suspension was placed on a freshly cleaved mica surface and allowed to dry at 80 °C overnight. In AFM, the topography and phase images of 0.1 % aqueous suspensions were recorded simultaneously in tapping mode at ambient conditions using commercial Si cantilevers from MicroMasch.

The morphology of the nanocomposites was characterized using a Cambridge 360 scanning electron microscope (SEM), at an acceleration voltage 10 kV. The samples were sputter coated with Au to avoid charging. The morphology of the nanocomposites was studied using a HITACHI S-4300 SE, field emission SEM at high vacuum and an acceleration voltage of 10 kV. The samples were fractured in liquid nitrogen and sputter coated with Au/Pd alloy to avoid charging.

X-ray Diffraction

A Siemens Diffractometer D5005 wide angled X-ray diffraction (WAXD), was used to study the crystallinity of the cellulose nanoreinforcements separated from the wood sources. The samples were exposed for a period of 11 s for each angle of incidence (δ) using a Cu X-ray source with a wavelength (λ) of 1.541 Å. The angle of incidence was varied from 1.5 to 40 by steps of 0.06 °.

Thermal and Mechanical Properties

The effect of the temperature on tan δ and storage modulus of prepared materials was studied using a DMTA V (Rheometrics Scientific). DMTA was run in tensile mode with thin film specimens, ~0.4 mm thick, and the gap was 15mm. The heating rate, strain rate, and frequency used were 3 °C/min, 0.1 % and 1 Hz respectively.

The mechanical strength and modulus of the composites were analyzed with a Sintech-1 machine model 3397-36 in tensile mode with a load cell of 50 lb using ASTM D 638 procedure. The extruded samples were compression molded into films approximately 0.5 mm thick at 180 °C. The specimens were cut into a dumbbell shape with a ASTM D 638 type V die. Tensile tests were performed at a crosshead speed of 2.5 mm/min. The values reported in this work are the average of at least five measurements.

Results and Discussion

The AFM and TEM pictures of cellulose whiskers obtained after the acid hydrolysis of MCC are shown in Figure 2. The cellulose whiskers were isolated into individual crystallites and dispersed uniformly in the aqueous medium having a needle shaped structures. These images show a large number of individual cellulose crystallites with an average diameter of 10-15 nm and length of 300-400 nm. The average aspect ratio (l/d) for the whiskers was estimated to be 40.

The structure of MFs characterized by TEM and AFM is shown in Figure 3. Figure 3a and 3b both indicate that at 0.1 % consistency, the MFs appear to be an entangled network of nano to microscale fibers less than 1 μm in diameter uniformly dispersed in aqueous medium.

The structure and crystallinity in the cellulose microfibers and whiskers were studied using X-ray diffraction. The diffraction patterns are shown in Figure 4. In both curves, the peaks are observed at 2θ = 15.2, 16.5, 22.8°, corresponding to a Cellulose I structure. The intensity of the peaks was higher for cellulose whiskers samples, showing that cellulose whiskers are more crystalline than microfibers.

Figure 5 shows SEM images of PLA cellulose nanocomposites containing 5 wt% of cellulose nanowhiskers/microfibers and 5 wt% PEG. Figure 5a is the overview of PLA-PEG-CNW composite, and the micrograph shows CNW in the PLA matrix. It is possible to see both isolated whiskers and agglomerates and that the dispersion is not uniform. Figure 5b shows a more detailed view of this composite. However, the diameters of the whiskers were observed to be much

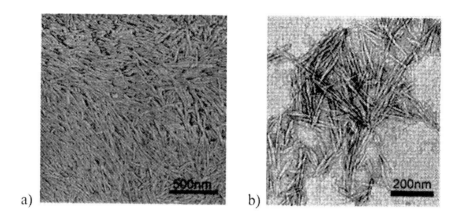

Figure 2. Micrograph of diluted aqueous suspension of cellulose nano whiskers a) AFM (0.1 %), b) TEM (0.01 %).

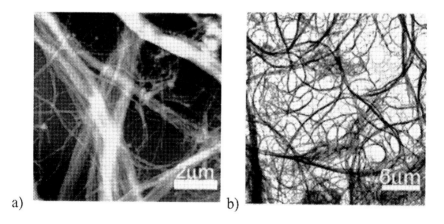

Figure 3. Micro graph of diluted aqueous suspension of microfibrillated cellulose, a) AFM and b) TEM

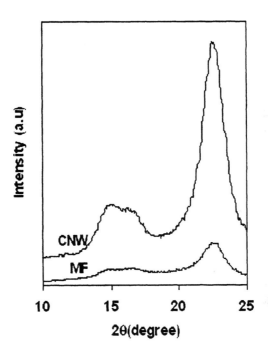

Figure 4. WAXD pattern of freeze dried cellulose nanowhiskers(CNW) and cellulose microfibers (MF).

124

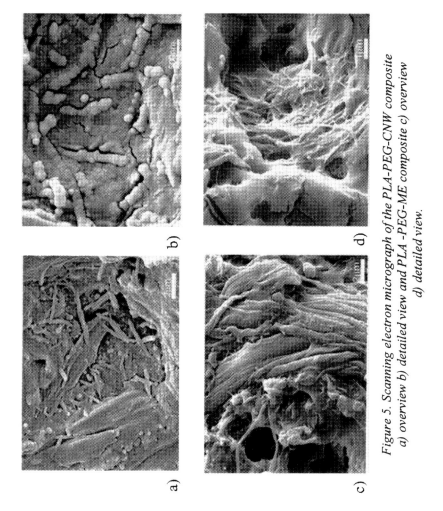

Figure 5. Scanning electron micrograph of the PLA-PEG-CNW composite a) overview b) detailed view and PLA -PEG-ME composite c) overview d) detailed view.

higher than the original whiskers introduced into the system. Figure 5c shows an overview of PLA-PEG-MF composite. Here we can see MF of different diameters up to ~1 μm embedded in the matrix. Figure 5d is a more detailed view of entangled networks of MF embedded in the matrix. The micrographs of the composites without PEG are not shown as the close observation of these composites showed more microscale particles compared to the sample containing PEG. The conclusion from the microscopy is that all the composites contain reinforcements dispersed in both nano- and microscale, and the dispersion / separation is better in the presence of PEG.

Figure 6 shows the viscoelastic behavior of PLA and cellulose nanocomposites. The storage modulus as function of temperature, in Figure 6a, shows that the storage modulus is slightly higher for the PLA-MF nanocomposite compared to pure PLA over the entire temperature scan. That is not the case for PLA-CNW composite, the CNW improve the storage modulus up to 60°C. Figure 6b, the plot of tan δ as a function of temperature, shows that the tan δ peak temperature has shifted slightly to higher temperatures for the PLA-MF composite (~70°C), compared to pure PLA (~68°C). The shift of tan δ peak temperature means an interaction between PLA and cellulose microfibers and can indicate that the molecular chains of PLA are to some extent restricted because of the microfibers. The addition of CNW seems not to affect the tan δ peak temperature. It is also observed that the intensity of the tan δ peaks are higher for both nanocomposites compared to pure PLA. Usually, the addition of fibres or other additives will decrease the tan δ peak intensity due to a reduction of material taking part in the relaxation process. Therefore increased tan δ intensity is an unexpected and uncommon behavior, and is an indication that the PLA composites are more amorphous than pure PLA. The tested materials were prepared in the same way, having similar cooling route in room temperature and the nano whiskers and microfibers was expected to act a nucleating sites for crystal growth in PLA and therefore rather increase the crystallinity of the PLA composites than decrease it.

Figure 7 shows the viscoelastic behavior of PLA and the nanocomposites containing PEG. Figure 7a shows that the storage modulus is slightly higher for the both nanocomposites over the entire temperature scan and 7b) the tan δ peak temperatures were shifted slightly in both composites compared to PLA-PEG. But this increase in tan δ peak is very little and not significant.

On comparing these two systems (with and without PEG), it is possible to see that the addition of PEG had a slightly negative effect on the storage modulus as well on the tan δ peak temperature for PLA but the addition of CNW and MF improved both the modulus and tan δ peak. The results also showed that the magnitude of the storage modulus drop and the intensity of tan δ peak indicates that pure PLA as well as all the nanocomposite materials studied are highly amorphous in nature.

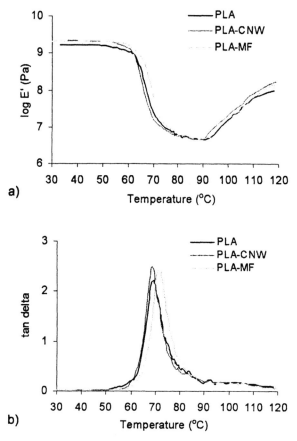

*Figure 6. DMTA curves of the nanocomposites
a) storage modulus b) tan δ.*

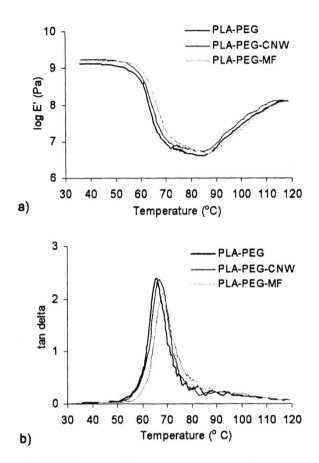

*Figure 7. DMTA curves of the nanocomposites with processing aid
a) storage modulus b) tan δ.*

Table IV shows the mechanical properties of prepared materials. The tensile strength of the composites containing CNW and MF did not show any significant differences compared to pure PLA. The stiffness of the nanocomposites without PEG showed an improvement compared to pure PLA. PEG, which was used as a processing aid, significantly improved the elongation to break for the pure PLA and also for the composites. Further more it caused a slightly decrease in tensile strength but did not affect the modulus. This is a positive effect for PLA because it is brittle polymer and usually additives which will improve the elongation to break will cause a significant decrease in the material stiffness.

Toughness, is the energy consumed by the samples before fracture and was calculated by integrating the area under the load-elongation curve. For comparison, the value was normalized per unit cross sectional area of the sample to express toughness as energy per unit area (kJ/m^2). The toughness PLA-PEG is not shown in Table IV because the samples had very high elongation without breaking, and therefore, it was not possible to calculate the energy. Similarly, the PLA-PEG-CNW system demonstrated high toughness values due to high sample elongation at break. In summary, the results from mechanical tests did not show any significant improvements in mechanical properties with S wt% cellulose nano reinforcement. The main reason for that might be an insufficient dispersion of the reinforcements in PLA, and formation of agglomerates.

Conclusions

This study was an attempt to prepare cellulose nanocomposites by melt extrusion technique using PLA as matrix. As the first step in this direction, cellulose nanowhiskers (CNW) and cellulose microfibers (MF) were isolated from commercially available wood sources viz. microcrystalline cellulose and wood pulp. It was found that CNW can be isolated from MCC by acid hydrolysis, and microfibers MF can be separated from wood pulp by cryocrushing and filtration. The cellulose whiskers and microfibers were characterized using microscopy and X-ray studies. AFM and TEM studies on deposited aqueous dispersions of whiskers showed that CNW exists as individual needle shaped crystals having a width of 10-15 nm and an average aspect ratio (l/d) of ~40. Similarly, characterization of MF from aqueous dispersions showed that at a concentration of 0.1 %, MF exist as entangled networks, having diameter up to 1 μm. The WAXD studies showed that the diffraction peaks of cellulose for CNW were much more intense than the corresponding peaks for MF, indicating that CNW is more crystalline compared to MF.

Table IV. Mechanical Properties of PLA and its Composites

Materials	Tensile Strength (MPa)	E-Modulus (GPa)	Elongation (%)	Toughness (kJ/m^2)
PLA	58 ± 6	2.0 ± 0.2	4.2 ± 0.6	35 ± 8
PLA-MF	58 ± 5	2.6 ± 0.1	2.8 ± 0.5	24 ± 8
PLA-CNW	57 ± 2	2.4 ± 0.3	3.3 ± 0.4	31 ± 7
PLA-PEG	51 ± 3	2.1 ± 0.6	> 50	-
PLA-PEG-MF	59 ±2	2.3 ± 0.1	3.3 ± 0.2	27 ± 2
PLA-PEG-CNW	47 ± 5	2.1 ± 0.3	5.4 ± 1.8	49 ± 19

Nanocomposites of PLA with 5 wt% CNW and MF were manufactured by twin screw extrusion. CNW and MF were fed into the extrusion by liquid pumping. This is a new way to prepare cellulose nanocomposites and this study indicates that water can be used as pumping medium for feeding nanosized additives to the polymer. However, fast evaporation of the water will lead to reaggregation of the whiskers and microfibers in the matrix polymer. This lead to poor dispersion in the nanolevel and therefore, it is advisable to use feeding medium which will provide more controlled evaporation during the processing.

Dynamic mechanical thermal analysis showed that nanocomposites had slight improvement in storage modulus and thermal stability compared to pure PLA. The tensile tests showed that the addition of cellulose whiskers and microfibers slightly improved the composite modulus while the strength was unaffected. The addition of PEG had a negative effect on the strength especially of pure PLA and also on CNW composites, Further PEG decreased the composites modulus but had a large effect on the elongation to break, especially for pure PLA and CNW composites. The toughness was improved for PLA-PEG-CNW system compared to other composite systems. In general, no significant improvement in mechanical properties was noticed for the nanocomposites compared to PLA. The main reason for the unexpected low change in mechanical properties is the non-uniform dispersion of the reinforcements in the PLA matrix and the generation of microscale agglomerates. Therefore, the next step in this direction is to improve the system by dispersing the reinforcements effectively in the matrix, using compatibilizers, surfactants and optimizing the feeding of the whiskers and microfibers into the extruder.

Acknowledgements

The authors would like to thank Cargill Dow LLC, Minnetonka, USA for supplied Nature Works™ PLA polymer and J. Rettenmaier & SÖHNE GmbH + CO, Rosenberg, Germany for supplied microcrystalline cellulose. We also thank the following researchers at NTNU, Norway: Dr. Bjørn Steinar Tanem for the TEM and AFM images, Ingvild Kvien, PhD student, for X-Ray (WAXD) work, and Daniel Bondeson and Magnus Bengtsson, PhD students, for their help with processing.

References

1. Krassig, H. A. *Cellulose: Structure, Accessibility and Reactivity, Polymer Monographs,* Gordon and Breach Science Publishers: Yverdon. **1993**; Vol 11, p. 376.
2. Hamad, W. *Cellulosic Materials, Fibers, Networks and Composites;* Kluwer Academic Publishers: The Netherlands, **2002**; p 47.
3. Bledzki A. K.; Gassan. J. *Prog. Polym. Sci.* **1999**, *24,* 221-274.
4. Eichhorn, S. J.; Baillie, C. A.; Zafeiropoulos, N.; Mwaikambo, L. Y.; Ansell, M. P.; Dufresne, A.; Entwistle, K. M.; Herrera-Franco, P. J.; Escamilla, G. C.; Groom, L.; Hughes, M.; Hill, C.; Rials, T. G.; Wild, P. M. *J. Mater. Sci.* **2001**, *36,* 2107-2131.
5. Mohanty, A. K.; Wibowo, A.; Misra, M.; Drzal, L. T. *Composites, Part A: Appl. Sci. Manuf* **2004**, *35,* 363-377.
6. Gindl, W.; Keckes, J. *Comp. Sci. Tech.* **2004**, *64,* 2407-2413.
7. Averous, L.; Fringant, C.; Moro, L. *Polymer.* **2001**, *42,* 6565-6572.
8. Oksman, K.; Skrifvars, M.; Selin, J-F. *Comp. Sci. Tech.* **2003**, *63,* 1317-1324.
9. Dufresne, A.; Dupreye, D.; Paillet, M. *J. Appl. Polym. Sci.* **2003**, 87, 1302-1315.
10. Favier, V.; Chanzy, H.; Cavaille, J. *Macromolecules.* **1995**, *28,* 6365-6367.
11. Angles, M. N.; Dufresne, A. *Macromolecules,* **2000**, *33,* 8344-8353.
12. Morin, A; Dufresne, A. *Macromolecules.* **2002**, *35,* 2190-2199.
13. Dufresne, A.; Cavaille, J. Y.; Vignon, M. R. *J. Appl. Polym. Sci.* **1997**, *64,* 1185-1194.
14. Dufresne, A.; Dupreye. D.; Vignon, M. R. *J. Appl. Polym. Sci.* **2000**, *76,* 2080-2092.
15. Samir, M.A.S.A., Alloin, F.. Paillet, M.; Dufresne, A. *Macromolecules.* **2004**, *37,* 4313-4316.
16. Nakagaito, A.N.; Yano, H. *Appl. Phys. A.* **2004**, *78,* 547-552.
17. Nakagaito, A.N.; Yano, H. *Appl. Phys. A.* **2005**, *80,* 155-159.
18. Grunert, M.; Winter, W. T. *J. Polym. Environ.* **2002**, *10,* 27-33.

19. Wu, Q.; Liu, X.; Berglund, L. A. *Sustainable Natural and Polymeric Composites-Science and Technology;* Eds.; 23rd Risø International Symposium on Materials Science Liholt, H.; Madsen, B.; Toftegaard, H. L.; Cendre, E.; Megnis, M.; Mikkelsen, L. P.; Sørensen, B. F., Eds.; Risø National Laboratory: Denmark, 2002; pp 107-113.
20. Lunt, J. *Polym Deg. Stab.* **1998**, *59*, 14.
21. Jamshidi, K.; Hyon, S. H.; Ikada, Y. *Polymer.* **1988**, *29*, 2229- 2234.
22. Oksman, K.; Mathew, A. P.; Bondeson, D.; Kvien, I. submitted to *Comp. Sci. Tech.*
23. Chakraborty, A.; Sain, M.; Kortschot, M. *Holzforschung.* **2005**, *59*, 102-107

Chapter 10

Preparation and Properties of Biopolymer-Based Nanocomposite Films Using Microcrystalline Cellulose

L. Petersson and K. Oksman

Department of Engineering Design and Materials, Norwegian University of Science and Technology, Rich Birkelands vei 2b, NO–7491, Trondheim, Norway

This chapter will give an idea of how the addition of microcrystalline cellulose (MCC) will affect the transparency, mechanical, thermal and barrier properties of biopolymers. It will also provide information on the production and structure of these nanocomposites. Two biopolymer matrices were used in this study, poly(lactic acid) (PLA) and cellulose acetate butyrate (CAB). The nanocomposite films were prepared by adding 5 wt % of MCC into both PLA and CAB using solution casting. Tensile testing showed increased strength (12 %) of the PLA nanocomposite and increased strength (30 %), toughness (300 %) and elongation to break (135 %) of the CAB nanocomposite. Dynamic mechanical thermal analysis (DMTA) also showed an improvement in storage modulus over the entire temperature range for the PLA nanocomposite. The barrier properties against UV/Visual light were improved, while the barrier properties against oxygen were reduced due to the addition of MCC.

Introduction

Biopolymers have received an increased interest lately due to more environmentally aware consumers, increased price of crude oil and global warming. These polymers are naturally occurring polymers that are found in all living organisms. Biopolymers are tailor made for specific functions within these organisms which makes them all unique. Some polymers are structural materials, while others are adhesives, energy stores, membranes or catalysts. Commercial biopolymers are categorized into three major groups (*1*). The first group is made up of biopolymers directly extracted from biomass, such as starch or cellulose (*1*). The second group is made up of polymers produced synthetically from biobased monomers, like poly(lactic acid) (*1*). The third and last group is made up of polymers produced by microorganisms or bacteria, like poly(hydroxyl alkanoates) (*1*). Biopolymers all come from renewable sources and are biodegradable, as a result their use will have a less negative effect on our environment compared to petroleum based materials. Today, biopolymers are used in a variety of applications, like therapeutic aids, medicines, coatings, food products and packaging materials.

There are a few biopolymers for engineering applications available, cellulose esters (CA, CAB, CAP), poly (hydroxy alkanoates) (PHAs), and polylactic acid (PLA). These polymers are competing to replace petroleum based polymers in a variety of applications (*2-4*). In order for these biopolymers to become a natural choice in plastic products they need to have acceptable prices, equally good performance as petroleum based polymers and enable the same production methods as petroleum based polymers. Studies have been carried out to compare the performance of biopolymers with petroleum based polymers in packaging applications (*4,5*). Petersen et al. have compared films made of different biodegradable polymers with films made of PE, PP and PS (*4*). They concluded that biodegradable polymers have high potential in packaging of highly respiring food, like fruits and vegetables, due to their high O_2CO_2 permeability ratio. However, they found that the major drawback with using biopolymers as packaging materials is their high water vapor permeability (*4*). Low thermal stability is another limitation to the use of biopolymers. PLA has for example a glass transition temperature (T_g) of around 60 °C, which can cause problems in packaging applications (*6*). Preparation of nanocomposites has been considered a promising method to improve the water vapor permeability together with the mechanical properties and thermal stability without affecting the transparency of biopolymers (*7,8*).

Polymer nanocomposites are produced by incorporating materials that have one or more dimensions on the nanometer scale (< 100 nm) into a polymer matrix. These nano materials are in the literature referred to as for example nanofillers, nanoparticles, nanoscale building blocks or nanoreinforcements. Nanocomposites have improved stiffness, strength, toughness, thermal stability, barrier properties and flame retardancy compared to the pure polymer matrix (8). Nanoreinforcements are also unique in that they will not affect the clarity of the polymer matrix. They will appear transparent since these nano materials are smaller than the wavelength of visual light (8). Only a few percentages of these nano materials are normally incorporated (1 – 5 %) into the polymer and the improvement is vast due to their large degree of surface area. The first research on polymer based nanocomposites was made by researchers at Toyota in Japan in the early 1990s (9).

When producing nanocomposites with biopolymer matrices it is important to use nanoreinforcements that ensure that the final material is biodegradable and solely based on renewable resources. Layered silicates are the most commonly used nano materials in the plastic industry today. These materials are naturally occurring materials, however they are not renewable or biodegradable. Cellulose on the other hand is abundant in nature, renewable, biodegradable and has good mechanical properties (10). It is therefore an interesting material to consider as nanoreinforcement. There are two different types of nanoreinforcements that can be obtained from cellulose fibers, microfibrils and nano whiskers. In plants or animals the cellulose chains are synthesized to form microfibrils. These are bundles of molecules that are elongated and stabilized through hydrogen bonding. One microfibril contains multiple elementary fibrils which are made up of multiple cellulose chains. These elementary fibrils have a diameter of around 2 to 20 nm depending on its origin (11). The elementary fibril is made up of amorphous and crystalline parts. The crystalline parts can be isolated by various treatments and these are the cellulose nano whiskers. As early as fifty years ago Rånby reported a process using acid hydrolysis that could separate cellulose whiskers from cellulose elementary fibrils (12). It is thought that these whiskers have mechanical strength that corresponds to the binding forces of neighboring atoms (11). As a result cellulose whiskers have far better mechanical properties than a majority of the commonly used reinforcement materials. During the 1990s the research in this area accelerated thanks to a group at CERMAV-CNRS in France. Today, it has been possible to remove cellulose nano whiskers from a few different sources like, tunicates (13-29) and wheat straw (21). These whiskers have then been incorporated into a number of different matrixes like for example, cellulose acetate butyrate (22), polyether (19), poly(hydroxyalkanoates) (15), poly(lactic acid) (23), poly(oxyethylene) (20), polyurethane (24), poly(vinyl chloride) (14), starch (16, 18) and silk fibroin (17).

Today cellulose nano whiskers are not commercially available. Microcrystalline cellulose (MCC) is a closely related product which is commercially available. MCC is made up of aggregated cellulose nano whiskers, see Figure 1. This product is for example used as a thickening agent in pharmaceuticals.

Cellulose nanocomposites can be produced using a few different processes, like for example solution casting, in-situ polymerization and melt intercalation (8). Solution casting is a small scale process mainly used in laboratories. This process enables the production of nanocomposites when only a very small amount of nanoreinforcement is available. In this process the nanoreinforcement is dispersed in a solvent in which the polymer is soluble. The dissolved polymer is then added to the solvent containing the nanoreinforcements and the mixture is poured into a container. When the solvent has been allowed to evaporate a polymer film is formed. The majority of the nanocomposites created with cellulose nano whiskers have been produced using this method (13, 15-18, 20, 22). In in-situ polymerization the nanoreinforcement is dispersed in a solution containing monomers. The polymerization of the polymer will take place in the solution surrounded by the nano materials. This method can be used both on a small and large scale. This method was used by Toyota to create their nylon nanocomposites (9) and has been used by Wu et al. to create cellulose nanocomposites (24). In melt intercalation the nanoreinforcement is blended with a melted polymer for example in an extruder. In this method the nanomaterial is not dispersed in a solution prior to blending, instead shear forces will try to disperse the nano materials evenly throughout the polymer. Melt intercalation is a large scale production method which therefore requires more materials compared to solution casting. Very few nanocomposites containing cellulose whiskers have been produced by using melt intercalation (14, 23).

The goal of the work presented in this chapter was to investigate the production, structure and properties of nanocomposites containing microcrystalline cellulose (MCC). The investigation was going to study the affect MCC would have on the transparency, mechanical, thermal and barrier properties of biopolymers. In this experiment MCC was swelled in order to allow polymer chains to penetrate the otherwise compact MCC particles. Two biopolymer matrices were used in this work, poly(lactic acid) (PLA) and cellulose acetate butyrate (CAB). The nanocomposite films were prepared by adding 5 wt % of swollen MCC into both PLA and CAB using solution casting. The structural details of the systems were studied using TEM. The mechanical performance was evaluated using conventional tensile testing and dynamic mechanical thermal analysis. The transparency and barrier properties of the films were investigated using UV-Vis spectroscopy and an oxygen permeability chamber. DSC was also carried out to explain the results from the oxygen permeability testing.

Experimental

Materials

Matrix. Poly (lactic acid) (PLA), Nature Works ^{TM} 4031 D, was supplied by Cargill Dow LLC, Minneapolis, USA. The material has a density of 1.25 g/cm^3, glass transition temperature (Tg) of 58 °C and melting point of 160 °C. The molecular weight (Mw) of the PLA is between 195,000 - 205,000 g/mol. Cellulose Acetate Butyrate (CAB), CAB-381-20 from Eastman Chemicals, USA was used. This CAB has a butyrate content of 37 wt %, an acetyl content of 13.5 wt % and a hydroxyl content of 1.8 wt %. The material has a density of 1.2 g/cm, glass transition temperature (Tg) of 141 °C, a melting point between 195 °C and 205 °C, and a molecular weight (Mw) of 70,000 g/mol.

Reinforcement. Microcrystalline Cellulose (MCC), Ceolus KG-802, was supplied by Asahi Kasei Corp., Tokyo, Japan. Ceolus KG-802 is commercially available and was used as a raw material for the swelling of cellulose nano whiskers (CNW).

Chemicals. Chloroform and Dimethylacetamide were purchased from Lab-Scan, Dublin, Ireland, and were used to swell and disperse the MCC. Lithium Chloride (LiCl) extra pure was purchased from Merck in Germany and was used during the swelling process of MCC. Silicon 100 from Novatio Europe N.V. was used to grease the Petri dishes prior to casting.

Sample Preparation

In this study MCC was swelled in the appropriate chemical for each biopolymer in order for the polymer chains to be able to penetrate the cellulose nano whiskers aggregated in the MCC particles.

The MCC used for PLA was swelled in distilled water. The MCC formed a 1 wt % solution which was stirred for 7 days at 60 °C. The solution was then subjected to 1.5 h sonification over 2 days in 10 min intervals in order to loosen up the MCC particles. The solution was then freeze dried and the MCC were redispersed in chloroform. The chloroform solution was sonified for 10 min at three different occasions during a 24 h period.

The MCC used for CAB was swelled in dimethylacetamide with a LiCl concentration of 0.5 wt %. The MCC formed a 1 wt % solution which was stirred for 7 days at 60 °C. The solution was then subjected to 1.5 h sonification over 2 days in 10 min intervals in order to loosen up the MCC particles before the solution was added to the dissolved polymer.

The nanocomposite films were prepared by solution casting. 15 wt % solutions of PLA in chloroform and CAB in DMAc were prepared by stirring on a hot plate at 50 °C until the pellets were fully dissolved (~5 h). The formulations, see Table I, were mixed and then sonified for 15 min prior to casting. The formulations were then casted in Petri dishes greased with silicon. The PLA formulations were left to evaporate in room temperature for a week, while the CAB formulations were first placed in an oven for 2 days at 60 °C and then allowed to rest at room temperature for 5 days. The prepared films had a thickness ≈0. 15 mm and a total dry weight of 5 g.

Table I. Prepared Formulations [wt %]

Material	PLA	CAB	MCC
PLA	100	-	-
PLA/S-MCC*	95	-	5
CAB	-	100	-
CAB/S-MCC	-	95	5

* S-MCC stands for swollen MCC

Characterization

The nano structure of the nanocomposite films was investigated using a Philips CM30 transmission electron microscope (TEM), at an acceleration voltage of 100 kV. A small sample with a cross-sectional area of 2x7 mm^2 was embedded in epoxy and cured overnight in room temperature. The final ultra microtoming was performed with a diamond knife at room temperature generating electron transparent foils, being approximately 50 nm in thickness. These foils were gathered on 300 mesh Cu grid.

Tensile testing was carried out using a miniature material tester Rheometric Scientific MiniMat 2000. The PLA samples were tested using the 1000 N load cell, while the CAB samples were tested using the 200 N load cell. The crosshead speed was the same for both materials, 2 mm/min. The samples were prepared by cuffing strips from the films with a width of 5 mm. The length between the grips was 15 mm. The results presented in the table are an average of 7 individual determinations, while the graphs are representative curves of the materials. The equipment used in this testing can only supply qualitative data, because the strain values are based on the rotational movement of the drive shaft.

Dynamic mechanical thermal Analysis (DMTA) was carried out on a Rheometric Scientific DMTA V in tensile mode. The measurements were carried out at a constant frequency of 1 Hz, a strain amplitude of 0.05 %, a heating rate of 3 °C/min and gap distance of 20 mm. The temperature range differed for the two different materials. PLA was tested between 15 °C – 100 °C and CAB was tested between 60 °C – 180 °C. All samples were placed in a vacuum oven at 24 °C for 4 days prior to testing in order to remove the remaining chemicals. The samples were prepared by cuffing strips from the films with a width of 5 mm. The results presented in the table are an average of 4 individual determinations, while the graphs are representative curves of the materials.

The permeability of oxygen was measured through the nanocomposite films using a home build testing chamber following the description given by Pye, D. G. et al (25). The initial oxygen pressure was 2.69 bar, room temperature [24 °C] was used and the area of the films exposed to the oxygen flow had a diameter of 36 mm.

Differential scanning calorimetry (DSC) was carried out on a Perkin Elmer DSC 7 using approximately 10 mg of the different materials. The PLA films were heated between 15 °C and 200 °C, while the CAB films were heated between 80 °C and 225 °C. A heating rate of 10 °C/min was used for all materials.

Transparency measurements were carried out on a Varian Inc. Cary 5 UV-Vis-NIR spectrophotometer. The X was varied between 600 and 200 nm and a spectral bandwidth of 2 was used together with a scan rate of 50 nm/min.

Results and Discussion

Material Structure

The TEM micrographs of the two nanocomposites are shown in Figure 2. These two images show that polymer chains have been able to penetrate the MCC structures and thereby creating nanocomposites. The swelling of the MCC

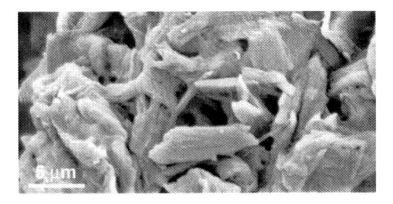

Figure 1. The structure of microcrystalline cellulose (MCC)

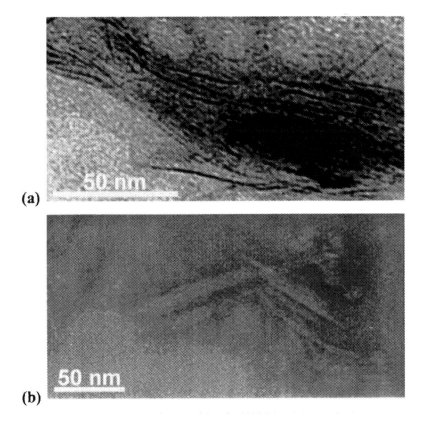

Figure 2. TEM analysis of the nanocomposite films.
(a) PLA/S-MCC, (b) CAB/S-MCC

prior to processing worked, because the MCC agglomerates still present in Figure 2 a and b are much smaller than the MCC particle presented in Figure 1.

The images in Figure 2 show that there are agglomerates still present in these two nanocomposite materials. By comparing Figure 2 a and 2 b, one can make the conclusion that DMAc with LiCl is more effective in separating the whiskers than water. This is not surprising since DMAc together with LiCl has been used to dissolve cellulose (26). The agglomeration of the whiskers will have an effect on the mechanical properties of the two nanocomposite materials. The properties of these films are most probably reduced compared to nanocomposite films with well dispersed cellulose whiskers. It should be noted that Figure 2 a might show a distortion of the PLA/S-MCC system, because the sample was unstable in the TEM examination and was destroyed shortly after the image was taken by the electron beam.

The processing method used in this experiment was not able to fully isolate the whiskers from the MCC. On the other hand this experiment shows that it is possible to use commercially available MCC without further acid hydrolysis to produce nanocomposite materials. Combining this processing method with a production process with large shear forces can maybe be enough to fully isolate the cellulose nano whiskers and produce a fully dispersed CNW nanocomposite material.

Mechanical Properties

The mechanical properties of the prepared nanocomposite films are presented in Table II. The nanocomposite materials showed an improvement in the tensile strength compared to the pure materials. Representative curves from the tensile testing can be seen in Figure 3. In more detail, the PLA/S-MCC nanocomposite showed a 12 % improvement in the tensile strength compared to the pure PLA. The CAB/S-MCC nanocomposite showed a 30 % increase in tensile strength and a 135 % increase in the elongation to break. It is interesting to see that the MCC is able to improve both the tensile strength and the elongation to break of the materials even tough the CNW are still not well separated and dispersed with in the matrix. The nanocomposite materials are not able to improve the E-modulus of the pure materials, see Table II. This is a result of the pore dispersion and the presence of larger agglomerates within the polymer matrices. The improvement in tensile strength is greater in the CAB system compared to the PLA system. This can be explained by that the whiskers were better dispersed in the CAB system compared to the PLA system, see

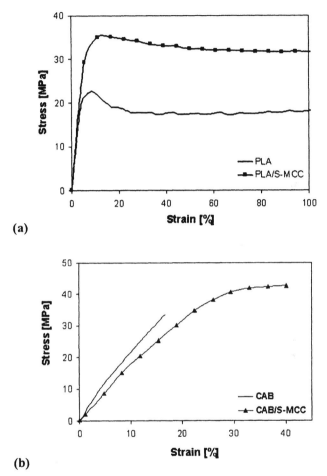

(a)

(b)

Figure 3. Stress-strain curves of the nanocomposite films.
(a) PLA/S-MCC, (b) CAB/S-MCC

Figure 2. The improvement in the mechanical properties will be greater when a fully dispersed system is created.

Table II. Average Mechanical Properties of the Nanocomposite Films

Composition	E-Modulus [GPa]		Strength [MPa]		Elongation to Break [%]
PLA	1.7	±0.2	28.5	±3.8	>100
PLA/S-MCC	1.5	+0.2	31.9	+2.8	>100
CAB	0.3	±0.1	30.3	±2.4	17
CAB/S-MCC	0.2	±0.0	39.5	±3.5	40

The toughness of the prepared films was also compared by determining the area beneath the stress-stain curves in Figure 3. The pure PLA had an elongation to break >> 100 % which resulted in a high toughness. The PLA/S-MCC material was able to maintain this high toughness due to the improvement in tensile strength. In the CAB system, the nanocomposite material was able to improve the toughness of the pure material with 300 % due to the increase in the elongation to break. This was part of our goal since CAB is known to be a brittle material.

It is important to remember when analyzing these results that solvent is still present in these polymer films due to that a vacuum oven was not used during processing. The amount of solvent left in the pure material and in the nanocomposites should be similar, since the two films were processed together using the same procedure. It is also important to remember that the testing equipment used to generate the mechanical properties can also not be used for quantitative analysis.

Dynamic Mechanical Properties

Values for the storage modulus (E′) and the tan δ peak of the materials are given in Table III. The storage modulus as a function of temperature is given in Figure 4. As can be seen in Figure 4 a, the storage modulus of the PLA/S-MCC film improved over a wide range of temperatures compared to the pure PLA. The improvement in storage modulus was most significant between 50 °C and 90 °C, where the molecular relaxation occurs for PLA. The storage modulus curve for PLA shows a drop in modulus after 45 °C and the curve later flattens out after 75 °C. For the nanocomposite film the drop in storage modulus is more

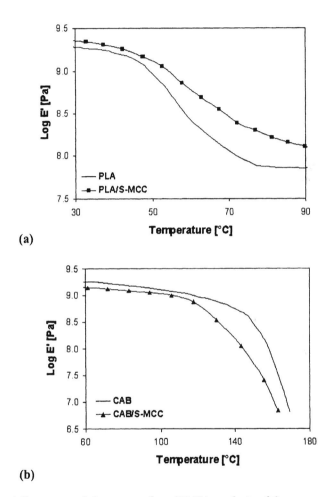

Figure 4. Storage modulus curves from DMTA analysis of the nanocomposite films. (a) PLA/S-MCC, (b) CAB/S-MCC

gradual and the curve flattens out at a higher temperature. Figure 4 a, indicates that the addition of the S-MCC was able to hinder both molecular and segmental motions of the PLA chains in the matrix. This can be explained by the small size of the whiskers which enables them to effect single polymer molecules. The PLA/S-MCC material is as a result a more thermally stable material, which was part of our goal since PLA has a low glass transition temperature (Tg) of around 58 °C.

Table III. Dynamic Mechanical Properties of the Nanocomposite Films

Composition	Low Temp E' [GPa]		High Temp E' [GPa]		Tan δ peak Temp. [°C]	
PLA	2.17	±0.2 [a]	0.13	±0.03 [c]	62	± 0
PLA/S-MCC	2.22	±0.2 [a]	0.24	±0.02 [c]	64	± 1
CAB	1.37	±0.3 [b]	0.31	±0.05 [d]	165	± 0
CAB/S-MCC	1.29	±0.1 [b]	0.05	±0.00 [d]	160	± 1

[a] 25°C, [b] 80°C, [c] 70°C, [d] 150°C

Unfortunately, the same result was not recorded for the CAB/S-MCC material, see Figure 4 b. The CAB/S-MCC material degraded during the DMTA run, all samples were brown when they were removed from the equipment. It is possible that this degradation was caused by the LiCl present in the system, since LiCl in DMAc is used to dissolve polysaccharides (26). Figure 4 b indicates that the degradation starts above a 100 °C and it should therefore most likely not have affected the results from the tensile testing. Figure 4 b also shows that the CAB/S-MCC material has a lower storage modulus that the pure CAB. This agrees with the tensile testing and indicates that the nanocomposite is a softer material compared to the pure CAB.

The storage modulus values received during DMTA analysis are higher than the values received during the tensile testing. This can be explained by the variation in strain of the two analysis methods. DMTA analysis tests the samples at low strain values, 0.05%, while larger strain values are tested during the tensile testing. Another explanation is the equipment used during tensile testing. This equipment is unable to record enough data points at low strain values to allow for modulus calculations in these areas. The increase in modulus for the DMTA analysis with about 1 GPA has also been recorded for other members of the group using the same two testing equipments (27).

The tan δ peak recorded for the PLA/S-MCC nanocomposite have been shifted to a slightly higher temperature compared to the tan δ peak for pure PLA, see Table III. It has been noted before that the tan δ peak broadens and shifts to higher temperatures for nanocomposites (8). This indicates that the nanoreinforcements have been able to reduce the segmental motions of the polymer matrix (8). This is can be explained by PLA chains penetrating the MCC structure and these PLA chains will therefore not be able to move as freely as the PLA chains in the matrix. Due to the degradation, the tan δ peak recorded for the CAB/S-MCC material shifts to a slightly lower temperature.

Oxygen Permeability

The results from the oxygen permeability test are shown in Table IV. These results show that the permeability of oxygen increased when the swollen MCC was incorporated into the two polymers. It is well known that other nanoreinforcements as for example layered silicates are able to reduce the permeability of different polymeric materials (8). This is often explained by viewing the layered silicate nanocomposite as a maze. The increased distance (tortuous path) the molecules have to travel through the material will have an effect on the barrier properties, since it will take a longer time for the molecules to pass through the material. The permeability properties are therefore said to be closely linked to factors that will influence the tortuous path like for example reinforcement loading, degree of exfoliation, orientation of reinforcements, shape of reinforcements, degree of crystallinity and porosity. The increase in permeability was larger for the PLA/S-MCC material than for the CAB/S-MCC material. This can partly be explained by the better dispersion of the cellulose whiskers in the CAB system.

Table IV. Oxygen Permeability Through the Nanocomposite Films

Composition	Permeability 10^{-7} $[m^3$ (STP) $*m/(m^2$ bar $h)]$
PLA	1.7
PLA/S-MCC	5.8
CAB	4.8
CAB/S-MCC	5.9

In order to better understand what caused the increase in permeability differential scanning calorimetry (DSC) was carried out to investigate if there was a difference in the degree of crystallinity between the pure polymers and the nanocomposites. The pure polymers had a higher degree of crystallinity compared to the nanocomposite materials. This can be established by comparing the heat of fusion of the different films, see Table V. Grunert and Winter have seen the same effect when they added untreated whisker to CAB (22). The degree of crystallinity will have an effect on permeability, since crystals are impenetrable by gas or liquids molecules. The difference in permeability is larger between the two PLA materials than between the two CAB materials. It is difficult to explain this large difference in permeability solely on crystallinity, because the difference in heat of fusion between the two PLA materials is smaller than the difference in heat of fusion between the two CAB materials. It is therefore possible that there are other mechanisms that also lead to an increase in oxygen permeability when adding MCC to PLA.

Table V. Average Differential Scanning Calorimetry Results of the Nanocomposite Films

Composition	Tm [°C]	Δhm [J/g]
PLA	167	29.0
PLA/S-MCC	167	26.3
CAB	161	15.3
CAB/S-MCC	160	10.2

UV-Visual Spectrometry

The results from the UV-Visual spectrometry scans are shown in Figure 5. There is a large reduction in the amount of light being transmitted through the nanocomposite films compared to the pure materials. The reduction is very similar for the two nanocomposite materials. The PLA/S-MCC material sees the largest reduction in the amount of light being transmitted trough the material. The PLA/S-MCC material lets 5.4 % of the light go through the material at 550 nm, which compares to 86.4 % for the pure PLA. The CAB/S-MCC system transmits 6.5 % of the light at 550 nm, which compares to 74.5 % for the pure CAB. The pure materials are letting more light through compared to the nanocomposite materials, which can partly be explained by the nanocomposite films not being completely transparent, see Figure 6. The nanocomposite films had a white haze, which was most probably caused by the MCC agglomerations present in the materials. It is difficult to determine if the reduction in light being transmitted through the nanocomposites is purely due to the white haze or if the

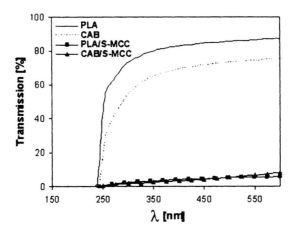

Figure 5. Transparency measurements of the nanocomposite films.

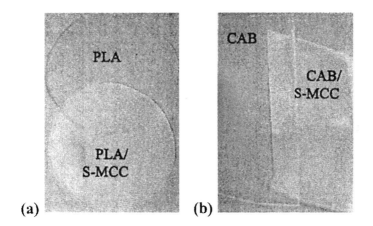

Figure 6. Optical clarity of the prepared films. (a) PLA/S-MCC,
(b) CAB/S-MCC

148

cellulose nano whiskers have been able to effectively block off the incoming light.

The pure materials had a higher degree of crystallinity compared to the nanocomposites, see Table V. The degree of crystallinity normally has an effect on the amount light being transmitted through a material, but not in this case. The white haze of the nanocomposite materials seems to have had a larger effect on the transmission of UV/Visual light than the difference in crystallinity between the pure and the nanocomposite materials.

Conclusions

The goal of the work presented in this chapter was to investigate the production, structure and properties of nanocomposites containing microcrystalline cellulose (MCC). The investigation was going to study the affect MCC would have on the transparency, mechanical, thermal and barrier properties of biopolymers. The two biopolymer matrices chosen for the study were poly(lactic acid) (PLA) which has a low thermal stability and cellulose acetate butyrate (CAB) which has a low elongation to break.

Nanocomposites were prepared by adding 5 wt % of swollen MCC to both CAB and PLA. The structure of the materials indicated that swelling the MCC prior to processing was enough to allow polymer chains to access the MCC particles. The morphology of the materials indicated that MCC was still present in aggregates, but these aggregates were smaller and looser than the original MCC particles.

The two nanocomposites were able to show improvements in tensile strength compared to the two pure materials. The tensile testing showed increased strength (12 %) of the PLA nanocomposite and increased strength (30 %), toughness (300 %) and elongation to break (135 %) of the CAB nanocomposite. The results from the dynamic mechanical thermal analysis (DMTA) also showed an improvement in storage modulus over the entire temperature range for the PLA nanocomposite. To conclude the mechanical testing, the thermal stability of the PLA was improved and the CAB nanocomposite was less brittle than the pure CAB. In short, by adding swollen MCC to these two biopolymers we were able to improve both of them even though their problems were very different.

The nanocomposite films were also investigated using UV-Vis spectroscopy and oxygen permeability in order to measure the performance of these materials in packaging applications. This is important since packaging applications is a commercial area of large interest to biopolymers. The results showed a large reduction in the amount of light being transmitted through the films, but unfortunately the barrier properties against oxygen were not improved.

It is believed that swelling commercially available MCC prior to processing can be combined with a production method using large shear forces to yield nanocomposite materials with fully isolated cellulose nano whiskers evenly distributed through out the polymer matrix.

Acknowledgements

The authors would like to thank Cargill Dow LLC, Minneapolis, USA for the supplied Nature Works™ PLA polymer, The Biopolymer Technology Department at Chalmers University of Technology, Göteborg, Sweden for the supplied CAB and Asahi Kasei Corp, Tokyo, Japan for the supplied microcrystalline cellulose. We would also like to thank Dr. B. S. Tanem for the transmission electron microscopy studies, Aji Mathew for great discussions, Elin Nilsen from the Department of Materials Technology at NTNU for help with the spectrometry studies and Arne Lindbråten together with Jon Arvid Lie from the Department of Chemical Engineering for help with O$_2$ permeability testing.

References

1. Petersen, K.; Nielsen, P. V.; Bertelsen, G.; Lawther, M.; Olsen, M. B.; Nilsson, N. H.; Mortensen, G. *Trends in Food Science & Technology* **1999**, *10*, 52-68.
2. Mohanthy, A. K.; Wibowo, A.; Misra, M.; Drzal, L. T.; *Composites: Part A* **2004**, *35*, 363-370.
3. Lunt J. *Polym. Degrad. Stab.* **1998**, *59*, 145-152.
4. Petersen, K.; Nielsen, P.; Olsen, M. *Starch* **2001**, *53*, 356-361.
5. Bastioli, C. *Starch* **2001**, *53*, 351-355.
6. Sinclair, R. G. *Polym. Mater. Sci. Eng.* **1995**, *72*, 133-135.
7. de Vlieger, J. J. In: *Novel food packaging techniques;* Ahvenainen, R., Ed.; Woodhead Publishing Limited: England, 2003; pp 519-534.
8. Alexandre, M.; Dubois, P. *Mater. Sci. Eng., R* **2000**, *28*, 1-63.
9. Usuki, A.; Kojima, Y.; Usuki, A.; Kawasumi, M.; Okada, A.; Fukushima, Y.; Kurauchi, T.; Kamigaito, O. *J. Mater. Res.* **1993**, *8*, 1185-1189.
10. Hamad, W. *Cellulosic Materials: Fibers, Networks and Composites;* Kluwer Academic Publishers; The Netherlands, 2002; p 47.

11. Samir, M. A. S. A.; Alloin, F.; Dufresne, A. *Biomacromolecules* **2005**, *6*, 612-626.
12. Rånby, B. *TAPPI* **1952**, *35*, 53-58.
13. Favier, V.; Chanzy, H.; Cavaille, J. Y. *Macromolecules* **1995**, *28*, 6365-6367.
14. Chazeau, L.; Cavaillé, J. Y.; Ganova, G.; Dendievel, R.; Boutherin, B. *J. Appl. Polym. Sci.* **1999**, *71*, 1797-1808.
15. Dufresne, A.; Kellerhals, M. B.; Witholt, B. *Macromolecules* **1999**, *32*, 7396-7401.
16. Anglès, M. N.; Dufresne, A. *Macromolecules* **2001**, *34*, 2921-2931.
17. Noishiki, Y.; Nishiyama, Y.; Wada, M.; Kuga, S.; Magoshi, J. *J. Appl. Polym. Sci.* **2002**, *86*, 3425-3429.
18. Mathew, A. P.; Dufresne, A. *Biomacromolecules* **2002**, *3*, 609-617.
19. Samir, M. A. S. A.; Alloin, F.; Sanchez, J.-Y.; Kissi, N. F.; Dufresne, A. *Macromolecules* **2004**, *37*, 1386-1393.
20. Samir, M. A. S. A.; Alloin, F.; Sanchez, J.-Y.; Dufresne, A. *Polymer* **2004**, *45*, 4149-4157.
21. Helbert, W.; Cavaillé, J. Y.; Dufresne, A. *Polym. Compos.* **1996**, *17*, 604-611.
22. Grunert, M.; Winter, W. T. *J. Polym. Environ.* **2002**, *10*, 27-30.
23. Oksman, K.; Mathew, A. P.; Bondeson, D.; Kvien, I. *Submitted to Compos. Sci. Technol.*
24. Wu, Q.; Liu, X.; Berglund, L. A. In: *Proceeding of the 23rd Risø International Symposium on Materials Science. Sustainable Natural and Polymeric Composites – Science and Technology;* Liholt, H.; Madsen, B.; Toftegaard, H. L.; Cendre, E.; Megnis, M.; Mikkelsen, L. P.; Sørensen, B. F., Eds.; Risø National Laboratory: Denmark, 2002; pp 107-113.
25. Pye, D. C.; Hoehn, H. H.; Panar, M. *J. Appl. Polym. Sci.* **1976**, *20*, 1921-1931.
26. Dupont, A.-L. *Polymer* **2003**, *44*, 4117-4126.
27. Bengtsson, M.; Gatenholm, P.; Oksman, K. *Compos. Sci. Technol.* **2005**, *65*, 1468-1479.

Chapter 11

Nanocomposites Based on Cellulose Microfibril

A. N. Nakagaito and H. Yano

Laboratory of Active Bio-Based Materials, Research Institute for Sustainable Humanosphere, Kyoto University, Kyoto, Japan

This chapter describes the production of nanocomposites exploiting the unusually good mechanical properties of cellulose microfibrils, from high-strength to optically transparent composites based on bacterial cellulose and microfibrillated cellulose from plant fibers. The strength were comparable to magnesium alloy (over 400 MPa) and the transparent composites exhibited coefficient of thermal expantion bellow 10 ppm/K.

Introduction

Cellulose is a polysaccharide polymer constituting the framework of higher plants and is the most abundant organic material on Earth. Among the primary sources of cellulose are wood fibers, which consist of hollow tubes made up of cellulose embedded in a matrix of hemicellulose and lignin. The tubular cell wall structure comprises a helically wound arrangement of cellulose microfibrils, nanofibers (4 nm x 4 nm) (*1*) consisting of semicrystalline cellulose chains parallel to their axes. As these molecular chains are extended and aligned, cellulose microfibrils possess a Young's modulus close to that of a perfect cellulose crystal, 138 GPa (*2*), with an estimated tensile strength well beyond 2 GPa. These mechanical properties are comparablé to those of aramid fibers, a well known high-strength synthetic fiber.

To exploit the unusually high strength of cellulose microfibrils, a cellulose morphology known as microfibrillated cellulose (MFC) was used as

152

reinforcement of thermosetting and thermoplastic matrices. This review aims to report the research being done in the Laboratory of Active Bio-based Materials during the last 5 years involving the development of nanocomposites based on cellulose microfibrils.

High-Strength Nanocomposites

Microfibrillated Cellulose Production

In the early 1980's, a new type of cellulose morphology was developed by Turbak *et al.* (*3*), known as microfibrillated cellulose or MFC. It is a form of expanded high-volume cellulose, moderately degraded and greatly expanded in surface area (Figure 1b), obtained through mechanical treatments of refining and high-pressure homogenization processes. Initially, this new material was intended to be used as additives in food, paints, cosmetics, and in medical products.

MFC was provided by Daicel Chemical Industries, Ltd., Japan and was obtained by a homogenizing process on kraft pulp consisting of Lodgepole Pine (*Pinus contorta*): 50%; White Spruce (*Pinus glauca*): 40%; and Douglas-fir (*Pinus menziesii*): 10%. First, a 3% concentration pulp fiber slurry was passed 30 times through a disk refiner and subsequently introduced into a device called an homogenizer. The slurry was pumped at high pressure through a spring-loaded valve assembly. The valve was opened and closed in a reciprocating motion, subjecting the cellulose to a large pressure drop of shearing and impact forces. This combination of repeated mechanical forces promoted fibrillation of the cellulose fibers and ultimately a high degree of microfibrillation. Cellulose fibers are split and unraveled to expose smaller fibrils and microfibrils, the latter having diameters in the range 10-100 nm (*5*).

Microfibrillated Cellulose-based Nanocomposites

Given the excellent mechanical properties of cellulose microfibrils and the nano-scale dimensions of MFC with an extensive reactive surface area, a great deal of reinforcement would be expected, once a good dispersion of fibrils had been attained. MFC-based nanocomposites were produced by lamination of MFC sheets using phenolic resin as the binder (*4, 6*). MFC was suspended in water at a fiber content of 0.2wt% and stirred for 48 hours. The suspension was vacuum filtered producing thin sheets about 0.1 mm in thickness which were oven dried at 70°C for 48 hours. In order to assure complete drying, they were further vacuum dried at 70°C for 5 hours and the weight was measured.

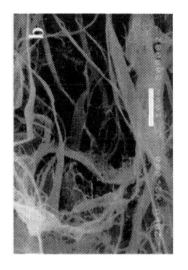
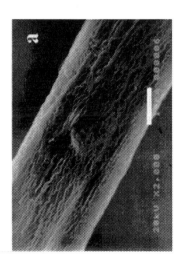

Figure 1. Scanning electron micrographs of: a) kraft pulp single elementary fiber; b) MFC. Scale bars: 10μm (Reproduced with permission from reference 4. Copyright 2005 Springer Science and Business Media).

The dried sheets of MFC were immersed in PF resin (PL-2340, Gun Ei Chemical Industry Co., Ltd., Japan) diluted in methanol in different concentrations in order to obtain various resin contents for the final composites. The immersed sheets were maintained in vacuum for 12 hours and kept at ambient pressure at 20°C over 96 hours. Impregnated mats were taken out of the solutions, air-dried for 48 hours, cut into smaller circles 5 cm in diameter, put in an oven at 50°C for 6 hours and weighed again. PF resin contents were calculated from the oven dry weights before and after impregnation. Finally the small circles were stacked in layers of about 25 sheets, put in a metal dye, and hot pressed at 160°C for 30 minutes at compressing pressures up to 100 MPa.

The mechanical properties were impressive, the Young's modulus achieved 19 GPa and the bending strength attained was about 370 MPa, figures comparable to those of commercial magnesium alloy. When compared to composites based on non-fibrillated pulp fibers (see Figure 2), MFC nanocomposites had slightly higher Young's modulus, but exhibited a bending strength around 1.5 times higher. Having similar modulus, the higher strength was a direct consequence of a higher strain at yield of MFC-based composites. The enhanced elongation results not only in higher strength, but also in higher toughness.

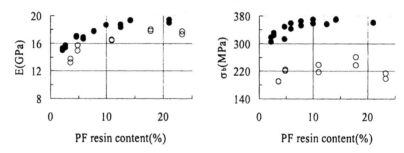

Figure 2. Comparison of mechanical properties (Young's modulus E and bending strength σ_b) of MFC-based(•) and non-fibrillated pulp-based (○) composites compressed at 100 MPa (Reproduced with permission from reference 4. Copyright 2005 Springer Science and Business Media).

The work of fracture is attributable to the highly extended surface area of networked nano-scalar fibrils which generates an increased bond density resulting in a crack delaying mechanism. As a consequence of the nano-scalar dimensions of the fibrils, fracture sites will be smaller and more widely distributed in the material volume. The nanqstructured material failure is therefore delayed and the strength is increased.

Effect of the Degree of Fibrillation of Cellulose on the Mechanical Properties of Nanocomposites

In order to determine how the degree of fibrillation of kraft pulp reinforcements affects the final composite's strengths, samples were produced using wood pulp with different levels of refining and homogenizing treatments (7). MFC is obtained by repeated mechanical action of a high-pressure homogenizer on wood pulp previously treated by a disk refiner. The number of passes through the homogenizer as well as the number of passes through the refiner determines the degree of fibrillation, resulting in different cellulose morphologies. The degree of fibrillation was evaluated indirectly by water retention values. Water retention is a physical characteristic related to the exposed surface area of cellulose (3) and serves as an approximate estimate of fibrillation. Phenolic resin was used again as the binder and the method to produce the composites followed the procedure described earlier.

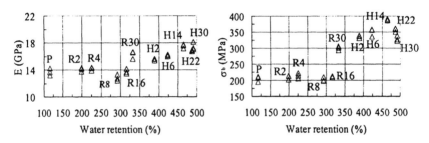

Figure 3. Young's modulus (E) and bending strength (σ_b) of nanocomposites vs. water retention of fibrillated kraft pulp with resin contents of 2.4~3.9 wt%. P corresponds to non-fibrillated pulp, plots labeled R relate to pulp treated by refiner only, and those labeled H refer to pulp additionally treated by homogenizer after 30 passes through the refiner. Numerals denote the number of passes through the refiner or homogenizer (Reproduced with permission from reference 7. Copyright 2005 Springer Science and Business Media).

Figure 3 shows the Young's modulus and bending strength as a function of the degree of fibrillation of pulp fibers, characterized as water retention values. There was no change in strength for composites prepared using pulp fibers treated by refiner up to 16 passes, however, a stepwise increase occurred when the treatment attained 30 passes through the refiner. The sudden increase in strength was most pronounced in the case of composites with resin contents below 5wt%, from about 200 MPa to 300 MPa, even though for composites with higher resin percentages the gap was smaller, it was still noticeable.

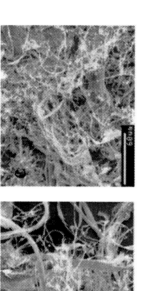
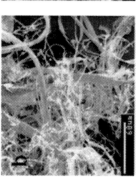
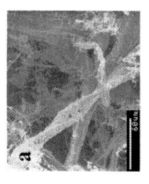

Figure 4. Scanning electron micrographs of: a) 8 passes through refiner pulp fibers; b) 16 passes through refiner pulp fibers; c) 30 passes through refiner pulp fibers. Scale bars: 60µm (Reproduced with permission from reference 7. Copyright 2005 Springer Science and Business Media).

Scanning electron microscopy observations, as shown in Figure 4, revealed that fibrillation of the fibers surface solely did not increase fiber interactions. Only the complete fibrillation of the bulk of the fibers resulted in an increment of mechanical properties, and additional fibrillation by homogenization treatment led to a linear increase of strength. Microfibrillation eliminates defects or weaker parts of the original fibers that would act as the starting point of cracks, but also increases interfibrillar bond densities creating a structure that favors ductility.

Effect of Fiber Content on the Thermomechanical Properties of Nanocomposites

The dependency of the reinforcing effect on the fiber content of nanocomposites is of considerable interest, especially considering that other than the enhanced modulus and strength, cellulose microfibrils are known to have extremely low coefficient of thermal expansion (CTE). It is estimated to be less than 0.1 ppm/K (8).

To obtain MFC-based composites with a broad range of fiber contents, the previously described method was modified to obtain MFC sheets and thermomechanical properties were measured (9). A dilute suspension of MFC in water was filtered, and the filtrate was freeze dried instead of oven drying. The freeze drying process avoids compaction through the sublimation of frozen water, producing MFC sheets with low density and high porosity, which enabled the fabrication of composites with a wider fiber content while maintaining the uniform fiber dispersion. Freeze dried sheets were impregnated with phenolic resin by immersion, air dried, and afterwards they were hot pressed at 160°C, adjusting the resin content by controlling the compressing pressure, not exceeding 10 MPa. The results of Young's modulus and CTE against fiber content are presented in Figure 5. The relationship was not linear, exhibiting a fast raise in Young's modulus at fiber contents up to 40wt% and lowering the increasing rate at higher contents. This tendency in reinforcement was also shown by the CTE measurements, the thermal expansion decreased rapidly, about 6 times from the CTE of PF resin to values similar to E-glass (10ppm/K) at fiber contents above 60wt%. From an atomistic view, Young's modulus and thermal expansion are both correlated to the depth of the atomic bond energy function.

The fast decrease in CTE can be attributed to the much larger Young's modulus of cellulose microfibril (138 GPa) compared to the PF resin (5 GPa) and to the extremely low CTE of cellulose microfibril (0.1 ppm/K). Hence, effective mechanical reinforcement as well as thermal expansion restraint could be attained at fiber contents from around 50wt% up, demonstrating the effective reinforcing capability of cellulose microfibrils.

158

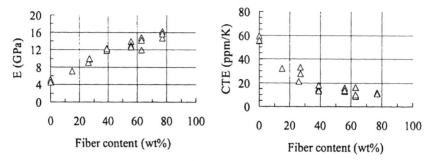

Figure 5. Young's modulus and CTE of MFC/PF resin composites as a function of fiber content.

Bacterial Cellulose Reinforced Nanocomposites

Apart from comprising the cell wall of plants, cellulose is also secreted by some bacterial species, as synthesized cellulose nanofibers called bacterial cellulose (BC). BC is produced by *Acetobacter* species cultivated in a culture medium containing carbon and nitrogen sources. It exhibits unique properties such as high mechanical strength and extremely fine and pure fiber networked structure. This structure is in the form of a pellicle made up of a random assembly of ribbon-shaped fibrils, less than 50 nm wide, which are composed of a bundle of much finer microfibrils, 2 to 4 nm in diameter (*10*). Instead of being obtained in a top-down process as with the fibrillation of fibers, BC is synthesized by bacteria in a reverse way, making a network of nanofibers somewhat straight, continuous, and dimensionally uniform. And as these nanofibers are made up of cellulose microfibrils, they possess enhanced mechanical properties. Nishi *et al.* (*11*) reported a strikingly high dynamic Young's modulus, close to 30 GPa, for sheets obtained from BC pellicles when adequately processed. Due to this remarkable modulus, BC sheets are ideal candidates as raw material to further enhance the Young's modulus of high-strength nanocomposites.

BC pellicles were boiled in a 1 wt% NaOH aqueous solution for 3 hours to remove bacterial cell debris. Pellicles consisting of 99% in volume of water were compressed under slight pressure to remove the excess water, and subsequently oven dried at 70°C for 48 hours. The sheets thus obtained were impregnated with the same phenolic resin and composites were prepared by lamination and hot pressing following the same procedure already described for the production of MFC-based composites.

Figure 6 shows the results for Young's modulus and bending strength of BC nanocomposites compared to MFC nanocomposites. The clearest difference between the two composites was in Young's modulus. BC composites revealed an exceptionally higher modulus than did MFC composites at any compressing pressure or resin content. However, the bending strength of BC composites, though slightly higher, was not as high in proportion to the Young's modulus. This discrepancy owes to the differences in strain at yield, BC composites have limited elongation up to fracture compared to MFC composites, as can be noted from the stress-strain curves of Figure 7.

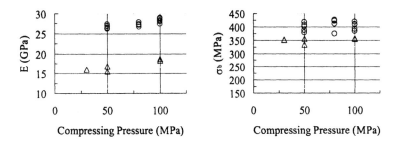

Figure 6. Young's modulus (E) and bending strength (σ_b) against compressing pressure of: ○ BC-based composites and △ MFC-based composites. The resin content of all BC composites was 12.4%; that of MFC composites was 10.3%~ 12.5% (Reproduced with permission from reference 12. Copyright 2005 Springer Science and Business Media).

According to Yamanaka *et al* (*13*), the high modulus of BC sheets could be attributed to a high planar orientation of the ribbon-like elements when compressed into sheets and to the ultra-fine structure of the elements, which allows more extensive hydrogen bonds. Another possible reason would be the relative straightness, continuity, and dimensional uniformity of the elements of BC. When composites were prepared using sheets obtained from fragmented BC pellicles, the mechanical properties and the deforming behavior up to fracture practically matched those of MFC composites. The change in micro-scale morphology most likely prevented the orientation of the ribbon-like elements when converted to sheets, in addition to loosing continuity and dimensional uniformity, causing the fragmented BC to have a structure similar to that of MFC sheets.

Biodegradable Green-composites Based on Microfibrillated Cellulose

Due to the unique characteristics of MFC, such as the vast surface area of the microfibrilar elements, it is possible to produce high-strength molded materials without the use of any kind of binder' (*14*). A high water content (90wt%) MFC creamy paste was set inside a metal die and by simultaneous slow compression and vacuum pumping, the excess water was reduced to

160

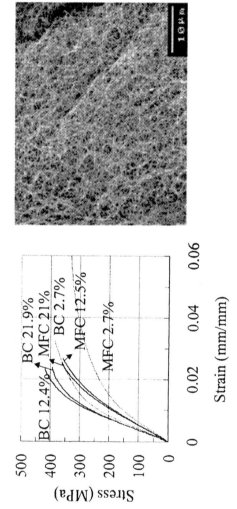

Figure 7. Stress-strain curves of MFC-based and BC-based composites. The percentages correspond to the resin content values. All samples were compressed at 100 MPa, except the BC 21.9%, which was compressed at 50 MPa. Scanning electron micrograph of a BC pellicle (Reproduced with permission from reference 12. Copyright 2005 Springer Science and Business Media).

around 50wt%, resulting in a molded sample. The sample was dried at 105°C to decrease water to about 2wt% and afterwards it was hot pressed at 150°C and under a pressure of 100 MPa. Microfibrillated cellulose with different degrees of fibrillation was used as the starting material, fibrillation levels were characterized by water retention values.

Modulus and strength increased linearly with increasing degree of fibrillation, and reached 16 GPa and 250 MPa, respectively, regardless of the fibrillation process. These values are approximately five times those for the original non-fibrillated pulp molded samples. High interactive forces seem to be developed between microfibrillated pulp fibers owing to their nanometer unit web-like network. In other words, the area of possible contact points per fiber are increased, so that more hydrogen bonds might be formed or the van der Waals forces are increased.

When 2wt% oxidized starch was added to the starting 14 passes through homogenizer-treated pulp and samples were obtained following the same procedure, the bending strength reached 320 MPa, although the Young's modulus decreased from 16 GPa to 12.5 GPa. These results indicate that the starch added might act not only as a binder but also as a plasticizer, so that the fibers became well packed, resulting in the formation of higher interactive forces between them.

To further extend the potential application of molded materials, the addition of thermoplastic agent was studied using a combination of microfibrillated pulp fibers and bio-degradable polylactic acid (PLA). A 14 passes through homogenizer-treated MFC and PLA fibers (length: 5mm, diameter: 12 μm) at a ratio of 7 to 3, were well mixed in a water suspension and filtered to make an approximately 20 μm thick dried sheet. Sheets were laminated and hot pressed at 170°C under 20 MPa and for 10 minutes to produce an approximately 1.5 mm thick specimen. During hot pressing, the laminated sheets were well bonded by thermal plasticization, and after cooling, the product became a plastic-like material. The modulus and bending strength of this plastic-like material were 17.5 GPa and 270 MPa, respectively.

The stress-strain curves on Figure 8 compares the mechanical properties of nanocomposites based on cellulose microfilbril with commercial conventional materials. It is clearly recognizable that MFC based composites have properties rivaling those of more traditional polymers and even metals, especially considering the specific strengths.

Optically Transparent Nanocomposites

BC-reinforced Optically Transparent Nanocomposites

A transparent nanocomposite reinforced with bacterial cellulose with high fiber content was reported by Yano *et al.* in 2005 (*15*). It was the first example

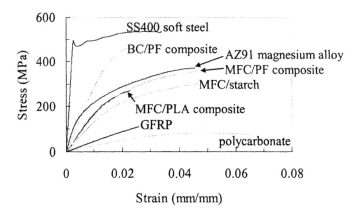

Figure 8. Stress strain curves of cellulose microfibril-based nanocomposites compared to conventional materials.

of an optically transparent composite at a fiber content as high as 70wt%, with a mechanical strength about five times that of engineered plastics, and the coefficient of thermal expansion similar to that of silicon crystal.

BC pellicles were compressed to remove the excess water and were afterwards dried at 70°C to completely remove the remaining water. Dried BC sheets were impregnated with neat acrylic resin (type UV3000A and type A600, Mitsubishi Chemical Corp., Japan), neat epoxy resin (type 2689, NTT Advanced Technology Corp., Japan), and a 20% methanol solution of phenol-formaldehyde resin (type PL-4414, Gun Ei Chemical Industry Co., Ltd., Japan) under vacuum for 12h. The latter PF is a transparent type, different from the previously used PF for making high-strength MFC composites. After impregnation, epoxy and acrylic resins were cured by UV light and phenolic resin impregnated sheets were hot-pressed at 150°C and 2 MPa for 10 min. Since the sheets before impregnation had a density of 1.0 g/cm^3 and considering that the density of cellulose microfibrils is 1.6 g/cm^3, the interstitial cavities of BC sheets accounted for about one-third of the sheet's volume. These cavities were filled with transparent thermosetting resins resulting in final composites with fiber contents between 60 to 70wt%.

Figure 9 shows the measurements of light transmittance versus wavelength of a BC/epoxy resin nanocomposite sheet, a BC sheet, and an epoxy resin sheet. In the wavelength interval from 500 to 800 nm, the BC/epoxy nanocomposite transmits more than 80% of the light, surface (Fresnel's) reflection included. Besides, when the transmittance of the BC/epoxy nanocomposite is compared with that of epoxy resin, the reduction in light transmission owing to the reinforcing nanofiber network is less than 10%. Even considering that for composite materials the transparency depends primarily on matching the refractive indexes of reinforcing elements and matrix, the BC reinforced nanocomposite seems to be less sensitive. The refractive index of cellulose is 1.618 along the fiber and 1.544 in the transverse direction whereas that of the impregnated epoxy resin is 1.522 at 587.6 nm and 23°C. Similarly, the refractive indexes of acrylic resins are 1.596 and 1.488 and of phenol-formaldehyde is 1.483, all at 587.6 nm and 23°C. The high transparency is due mainly to the nanosize effect, i.e., elements with size less than one-tenth of the wavelength prevents the scattering of light.

Another very attractive property of BC nanocomposites is the unusually reduced thermal expansion. The coefficient of thermal expansion (CTE) of the BC/epoxy combination was 6 ppm/K, an extremely low value compared to 120 ppm/K of the epoxy matrix. The CTE of BC/phenol-formaldehyde was even lower at 3 ppm/K, a figure as low as that of silicon crystal. The tensile strength measured reached values up to 325 MPa, with Young's modulus around 20 to 21 GPa. In addition, BC nanocomposites are light, flexible and easy to mold, making them promising candidate materials for a broad field of applications from flexible displays to windows of vehicles.

Optically Transparent Nanocomposites Reinforced with Plant Fiber-based Nanofibers

In order to fibrillate wood pulp fibers and reduce the size of fibril bundles to a greater extent, the refining/high-pressure homogenization process to obtain MFC was complemented by a grinding treatment (*16*). MFC obtained by 14 passes through the homogenizer has a wide distribution of fibrils width, from some tens of nanometers to some microns. As a result, impregnating these MFC sheets with acrylic resin produces transparent composites but not as much as BC/acrylic resin composites. The main reason for this reduced transparency might be attributable to light scattering caused by the larger fibril elements.

Since additional passes through the homogenizer did not improve the composites transparency, the 14 passes through homogenizer MFC was subjected to a grinder treatment, realized by 10 times iterations. It consists of a mechanical process that applies shearing stresses to the fibers through a commercial grinder fitted out with a pair of specially designed grinding disks. This additional treatment resulted in fibril bundles with dimensions 50 to 100 nm in width. This grinder-fibrillated MFC was dispersed in water at a fiber content of 0.2wt% stirring it for 24 hours. The suspension was vacuum filtered using polytetrafluoroethylene membrane filter (0.1-μm mesh), producing a thin sheet. Sheets were oven dried at 55°C for 48 hours and to assure a complete drying, they were further dried at 105°C for two hours, after which the weight was measured. Dried sheets were immersed in neat acrylic resin (TCDDMA, Mitsubishi Chemical Corp., Japan) and maintained in reduced pressure for 24 hours. Impregnated sheets were taken out and the resin was cured by UV light.

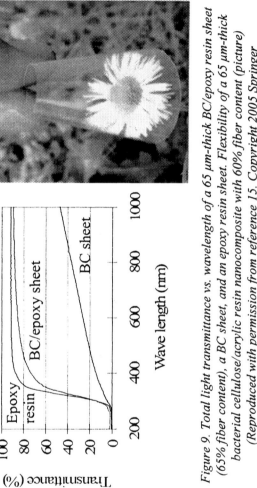

Figure 9. Total light transmittance vs. wavelength of a 65 μm-thick BC/epoxy resin sheet (65% fiber content), a BC sheet, and an epoxy resin sheet. Flexibility of a 65 μm-thick bacterial cellulose/acrylic resin nanocomposite with 60% fiber content (picture) (Reproduced with permission from reference 15. Copyright 2005 Springer Science and Business Media). (See page 1 of color inserts.)

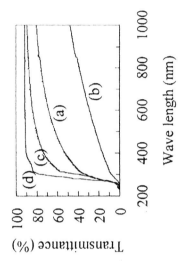

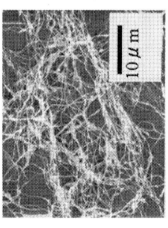

Figure 10. Scanning electron micrograph of 30-times grinder-fibrillated pulp (picture). Regular light transmittance of (a) 45 μm-thick grinder-fibrillated pulp/acrylic resin sheet, (b) 53 μm-thick high-pressure homogenizer-fibrillated pulp/acrylic resin sheet, (c) 60 μm-thick bacteria cellulose/acrylic resin sheet, and (d) 65 μm-thick acrylic resin sheet (Reproduced with permission from reference 16. Copyright 2005 Springer Science and Business Media).

The resulting increase in oven-dry weight during impregnation was used to estimate the fiber content in the final composites.

The results of light transmittance against wavelength of grinder fibrillated fiber/acrylic resin composite, compared to pure acrylic resin, BC/acrylic resin, and homogenizer fibrillated fiber/acrylic resin are presented in Figure 10. At the wavelength of 600 nm, grinder fibrillated fiber/acrylic composites with 70wt% fiber content, transmits 70% of light including surface (Fresnel's) reflection, what translates into a transmission reduction of just 20% compared to the transmittance of the pure acrylic resin. The light transmittance of homogenizer fibrillated fiber/acrylic composites at a 62wt% fiber content is 30%, or 60% degraded in relation to the pure resin.

While BC/acrylic resin exhibited the highest transmittance among all the composites, it is an indication that additional fibrillation of plant fibers might lead to higher light transmittances since the cellulose microfibril is 4nm wide and 4nm thick, still leaving room for improvements in the fibrillation process. The CTE of the grinder fibrillated fiber/acrylic resin was measured as 17 ppm/K whereas the CTE of the pure acrylic resin is 86 ppm/K, i.e., the grinder fibrillated nanofibers reduced the CTE of the matrix resin to one-fifth of its original value. The Young's modulus was 7 GPa, although it is lower than the modulus of BC composites, it means more flexibility, or they are easier to bend than BC composites.

Conclusion

The objective of these studies was to develop a new process to manufacture bio-nanocomposites exploiting the unusually unique properties of cellulose microfibrils. The concept is based on utilizing microfibrillated cellulose (MFC), a cellulose morphology obtained through microfibrillation of kraft pulp fiber by mechanical processes of refining and high-pressure homogenization. The treatment confers drastically large surface area to the material, characterized by a nano-scalar, web-like networked fibrils and microfibrils, which have the potential to increase interactive forces between reinforcing elements and matrix in a composite. Due to the nano-scalar dimensions of the fibrils, fracture sites in the composites will be smaller and more widely distributed in the material volume, delaying the nanostructured material failure and increasing its strength. The nanoscalar dimensions of microfibrillated cellulose not only provides mechanical reinforcement but also makes possible the reinforcement of transparent polymers with minimal light scattering by relying on the fact of having sizes far shorter than the wavelength of visible light. Cellulose microfibrils can reinforce transparent polymers providing strength, toughness, and reduced thermal expansion without hindering the optical transparency.

Microfibrillation of plant fibers is definitively one of the most prominent means to exploit the remarkable properties of the cellulose microfibril, which so far only nature can produce.

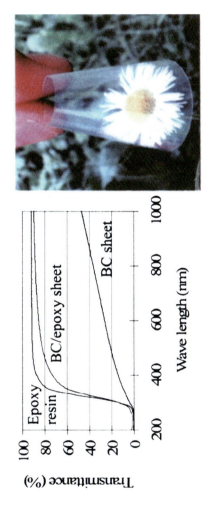

Figure 11.9. Total light transmittance vs. wavelength of a 65 µm-thick BC/epoxy resin sheet (65% fiber content), a BC sheet, and an epoxy resin sheet. Flexibility of a 65 µm-thick bacterial cellulose/acrylic resin nanocomposite with 60% fiber content (picture) (Reproduced with permission from reference 15. Copyright 2005 Springer Science and Business Media).

References

1. O'Sullivan, AC., *Cellulose: The structure slowly unravels.* Cellulose, 1997, *4*, 173-207.
2. Nishino, T., K. Takano, and K. Nakamae, *Elastic-Modulus of the Crystalline Regions of Cellulose Polymorphs.* Journal of Polymer Science Part B-Polymer Physics, **1995**, *33*, *11*, 1647-1651.
3. Turbak, A.F., F.W. Snyder, and K.R. Sandberg, *Microfibrillated cellulose, a new cellulose product: properties, uses, and commercial potential.* Journal of Applied Polymer Science: Applied Polymer Symposium, **1983**, *37*, 815-827.
4. Nakagaito, A.N. and H. Yano, *Novel high-strength biocomposites based on microfibrillated cellulose having nano-order-unit web-like network structure.* Applied Physics a-Materials Science & Processing, **2005**, *80*, *1*, 155-159.
5. Herrick, F.W., R. L. Casebier, J. K. Hamilton, and K. R. Sandberg, *Microfibrillated cellulose. morphology and accessibility.* Journal of Applied Polymer Science: Applied Polymer Symposium, **1983**, *37*, 797-813.
6. Nakagaito, A.N., H. Yano, and S. Kawai, *Production of high-strength composites using microfibrillated kraft pulp.* 6th Pacific Rim Bio-Based Composites Symposium, Portland 2002, **2002**, 171-176.
7. Nakagaito, AN. and H. Yano, *The effect of morphological changes from pulp fiber towards nano-scale fibrillated cellulose on the mechanical properties of high-strength plant fiber based composites.* Applied Physics a-Materials Science & Processing, **2004**, *78*, *4*, 547-552.
8. Nishino, T., I. Matsuda, and K. Hirao, *All-cellulose composite.* Macromolecules, **2004**, *37*, (*20*), 7683-7687.
9. Nakagaito, A.N., *Bio-nanocomposites based on cellulose microfibril.* PhD thesis, **2005**, 45-55.
10. Yamanaka, S., et al., *The Structure and Mechanical-Properties of Sheets Prepared from Bacterial Cellulose.* Journal of Materials Science, **1989**, *24*, *9*, 3141-3145.
11. Nishi, Y., M. Uryu, S. Yamanaka, K. Watanabe, N. Kitamura, M. Iguchi, and S. Mitsuhashi, *The Structure and Mechanical-Properties of Sheets Prepared from Bacterial Cellulose. 2. Improvement of the Mechanical-Properties of Sheets and Their Applicability to Diaphragms of Flectroacoustic Transducers.* Journal of Materials Science, **1990**, *25*, (*6*), 2997-3001.
12. Nakagaito, A.N., S. Iwamoto, and H. Yano, *Bacterial cellulose: the ultimate nano-scalar cellulose morphology'for the production of high-strength composites.* Applied Physics a-Materials Science & Processing, **2005**, *80*, (*1*), 93-97.

13. Yamanaka, S., M. Ishihara, and J. Sugiyama, *Structural modification of bacterial cellulose.* Cellulose, **2000**, *7*, (*3*), 213-225.
14. Yano, H. and S. Nakahara, *Bio-composites produced from plant microfiber bundles with a nanometer unit web-like network.* Journal of Materials Science, **2004**, *39*, (*5*), 1635-1638.
15. Yano, H., J. Sugiyama, A. N. Nakagaito, M. Nogi, T. Matsuura, M. Hikita, and K. Handa, *Optically transparent composites reinforced with networks of bacterial nanofibers.* Advanced Materials, **2005**, *17*, (*2*), 153-155.
16. Iwamoto, S., A. N. Nakagaito, H. Yano, and M. Nogi, *Optically transparent composites reinforced with plant fiber-based nanofibers.* Applied Physics a-Materials Science & Processing, **2005**, *81*, (*6*), 1109-1112.

Chapter 12

Cellulose Microfibers as Reinforcing Agents for Structural Materials

A. Chakraborty[1], M. Sain[1,2], and M. Kortschot[1]

[1]Department of Chemical Engineering and Applied Chemistry, University of Toronto, 200 College Street, Toronto, Ontario M5S 3E5, Canada
[2]Faculty of Forestry, University of Toronto, 33 Willcocks Street, Toronto, Ontario M5S 3B3, Canada

Since the use cellulose whiskers, cellulose microfibrils, and cellulose fibres of micro- and nano-scale diameters as reinforcing agents in composites is increasing rapidly, there is a need to review the microstructure of cellulose in great detail. In this paper, the concept of "microfibres" of cellulose has been developed based on a consideration of bleached kraft pulp cell wall morphology. First, the structure of a cellulose microfibril in wood cell wall has been elucidated through a structural analysis of wood cell wall. Subsequently, the need to define microfibres as "fibres" for structural applications has been demonstrated, and contrasted with the convential definition of microfibrils, which are attached to the cell wall at one end. It is proposed that microfibres should have a minimum aspect ratio of 50 for adequate stress transfer from the matrix to the fibre. Consequently, microfibres were defined as cellulose strands 0.1–1 µm in diameter, with a corresponding minimum length of 5–50 µm. The reinforcing potential of cellulose microfibres was demonstrated by using them in composites with a polyvinyl alcohol (PVA) matrix.

Introduction

Nanocomposites incorporating cellulose microfibrils are increasingly of interest in the composite science community. In this context, in view of an apparent lack of clarity in existing literature, there is a pressing need to clearly define the concept of microfibres of cellulose. Therefore, the objective of this paper is to formally define the term "microfibres" in terms of structural reinforcement capacity, in contrast to the conventional concept of "microfibrils".

Microfibrils in plant cells can be defined from the perspective of biogenesis. As noted by Mühlethaler (*1*), if cell walls of different origin are extracted with dilute acid or alkali, random or parallel fibrous textures become visible. This indicates that the non-cellulosic components such as pectin, hemicellulose and lignin are deposited between the cellulose fibrils. These fibrils are called microfibrils. In the biogenetic pathway of plant growth, a microfibril represents the unit structural component created during the deposition of cellulose in the cell wall. In other words, from the point of view of plant cell morphology, the microfibril is the basic structural unit produced during photosynthesis.

In this study, the structural hierarchy of cellulose is investigated in order to analyze the structure of a microfibril in a logical way. Subsequently, the concept of a fibre as a structural material is discussed. Finally, with these two concepts adequately elucidated, a formal definition of the term "cellulose microfibre" is proposed.

Cellulose Structure

Cellobiose Molecule

Cellulose is the most important constituent of the cell wall, and forms a framework around which all other cell wall components are deposited (*2*). α-cellulose, the fraction of the whole cell wall left after delignification and extraction of hemicelluloses and pectic substances, consists mostly of linear chains of β-1,4 linked glucose. It differs from the structure of α-glucose in the position of the OH group on carbon 1, as illustrated in Figure 1.

However, these (β-glucose molecules are not normally present in this native state. Glucose mostly forms a ring structure in which an oxygen bridge links carbon atoms number 1 and 5 shown in Fig. 1B. When two heterocyclic molecules condense with the loss of one molecule of water, the disaccharide, cellobiose, is formed. The structure of cellobiose is shown in Figure 2.

Further polymerization of glucose results in long chains of molecules. The continuous chain in a cellulose molecule is depicted in Figure 3. As the chemical structure shows, the hydroxyl groups of C-atoms, 2, 3 and 6 in cellulose are free to form hydrogen bonds with hydroxyls of adjacent chains, giving the cellulose

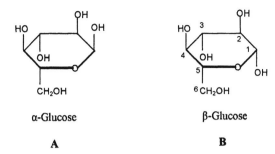

Figure 1. The structure of glucose molecule. A: α-glucose B. β-glucose.

a superstructure of considerable lateral order (*3*). This relatively weaker but numerous hydrogen bonds between molecules imparts immense strength to the polymer structure. Although there seems to be a good deal of dispute over the matter, particularly due to different biological sources of cellulose, the average number of units in a cellulose molecule may lie between 3,000 and 10,000, thus giving molecular lengths of between 1.5 and 5.0 microns (*2*). Rydholm (*3*), for example, reported an average degree of polymerization (DP) between 4,000 and 5,000 for cellulose in both softwood and hardwood. Ohad and Mejzler (*4*), however, mentioned that the average chain length may consist of as low as 2,500 monomers.

The chain length of each cellobiose molecule is 1.03 nm, and one molecule is separated laterally from others by a distance of 0.835 nm. The widely accepted structure of an elementary cell is a monoclinic one with the following dimensions:

$$a = 0.83 \text{ nm}$$
$$b = 1.03 \text{ nm}$$
$$c = 0.79 \text{ nm}$$
$$\beta = 84°$$

The hydroxyl oxygens of two adjacent chains are at a distance of only 0.25 nm in the direction of the a-axis, allowing complete hydrogen bonding (*3*).

Elementary Fibrils and Cellulose Whiskers

Several cellobiose units group together to form long thin crystallites called elementary fibrils, also known as micelles. These elementary fibrils are highly crystalline along the entire length, and there is no segmentation of crystalline and paracrystalline or amorphous regions. The elementary fibrils are separated laterally from each other by non-crystalline regions. By measuring the cross

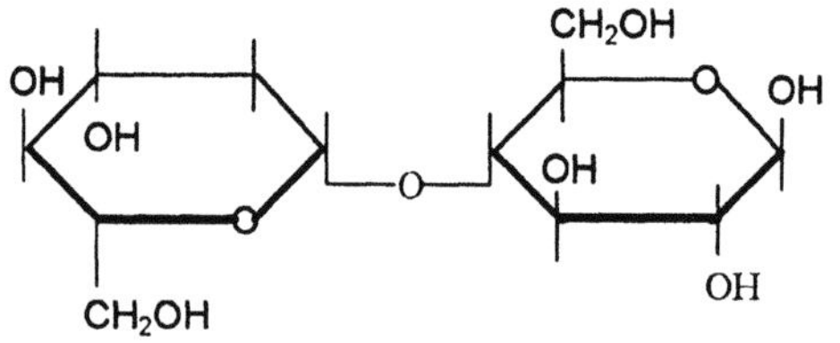

172

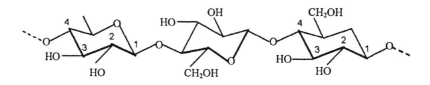

Figure 2. The structure of cellobiose molecule (β-1,4 Linked Glucose).

Figure 3. Part of a cellulose molecule – Cellulose β-1,4 Linked Glucose.

section of the elementary fibril, and knowing the space occupied by one cellulose molecule, the number of molecules in each thread can be calculated. Such a calculation was conducted by Frey-Wyssling and Mühlethaler (5), revealing that an average of 37 molecules are packed into one elementary fibril. For geometric reasons, it was assumed in this calculation that a cross section contains 7 molecules along the (101)-plane, and 6 molecules perpendicular to it.

The widths of these fibrils are reported to vary according to the source of the cellulose. Typical values are reported in the range between 3.5 nm (6) to 5 nm (2). Rydholm (3) has noted that these elementary fibrils are 10 nm wide and 3 nm thick. Ohad and Mejzler (4), on the other hand, observed a width of 3–3.5 nm and a thickness of 1.6-2 nm for these fibrils of rectangular cross sections. These crystallites are arranged in the shape of a ruler.

A cellulose molecule also has alternating regions of crystallinity along its length. Hence, the length of each crystalline zone is much smaller than the chain length of a cellulose molecule. On the basis of X-ray studies and microscopic imaging, it has been concluded that each crystallite region constitutes of at least 120 monomers, giving a minimum chain length of 60 nm (4) of the elementary fibrils.

In composite science, the term "cellulose whiskers" is commonly used. This refers to "zero-defect" crystalline regions in the cellulose chain, and therefore, are analogous to the elementary microfibrils in trees and plants. Therefore, elementary fibrils and whiskers refer to the same structural unit, the former nomenclature being used in botany, and the latter, in materials and composite science.

Microfibril Morphology Within the Cell Wall

The elementary fibrils, in turn, are organized in groups of up to 20 to form a microfibril (2). The width of microfibrils has been found by various researchers to lie within a range of 3.5 to 38 nm, depending in the source of cellulose (3,7,8). Specifically for wood microfibrils, Hodge and Wardrop (9) observed dimensions of 5 to 10 nm. From electron micrographs, Vogel (10) showed that microfibrils are composed of rectangular smaller units having dimensions of 3 X 10 nm.

Clowes and Juniper (2) defined microfibrils as "thin cellulose threads 8 x 30 nm in diameter, and up to 5 μm long comprising several elementary fibrils and forming the skeleton of higher plant cells."

A number of structures for the microfibril have been proposed. These models differ mainly in the presentation of the less ordered regions. All models can be reduced to one of three basic structures (11);

I. "the longitudinally arranged molecules change from one ordered region to the subsequent one, the transition areas being less ordered regions (fringed micelle system)"

174

II. "the fibrillar units are individual cords consisting of longitudinally arranged molecules and sequences of ordered and disordered regions"

III. "the ordered regions are packages of chains folded in a longitudinal direction, the areas containing the turns between adjacent chain packages being the less ordered regions."

The most commonly considered microfibril structure is that of a flat ribbon, having one transverse dimension four times than the other (12). This structure (illustrated in Figure 4), consists of several totally crystalline elementary fibrils (E), between and around which lie the less regularly arranged glucose chains, i.e., the paracrystalline region (P). This structure is widely considered to be the basic structural unit of cellulose. Linkages may also be made between the outside of the microfibril and non-cellulose polysaccharides (N).

Another less conventionally-accepted reconstruction of microfibrils has been postulated by Manley (13). He proposed that a microfibril consists of chains of cellulose arranged in a helical structure 3.5 nm wide. In this helix, the cellulose molecules run almost at right angles to the "tape" when unwound, but parallel to its axis when in the helical form. This structure of a microfibril is shown in Figure 5.

Rationale for Defining "Microfibres" for Structural Applications

In the manufacture of composites with cellulose fibres of submicron diameter as reinforcing agents, we consider an array of cellulose chains that act as a "fibre". Microfibrils, however, are all attached to the cell wall, and have never been considered as a separate entity. However, if cellulose fibrils (as opposed to entire wood fibres) are considered for use as reinforcing agents in polymer matrices, these microfibrils need to be separated from the cell wall, and this will result in structural components having a certain aspect ratio, i.e., length to diameter ratio. Once it has such an aspect ratio, they would form a fibre instead of a fibril. Hence, there is a need to define the term "microfibre", as opposed to a microfibril, as a discrete fibre.

In the literature discussed above, it is clear that microfibrils are not entirely crystalline in nature. Although microfibrils are generally considered to be devoid of non-cellulosic components, they do have intermittent amorphous and semicrystalline zones. Similar to lignin and hemicellulose, these amorphous regions would lower the mechanical strength of the fibril. Moreover, these non-crystalline zones are arranged arbitrarily within the microfibril chain. Therefore, microfibrils do not have uniformity in properties.

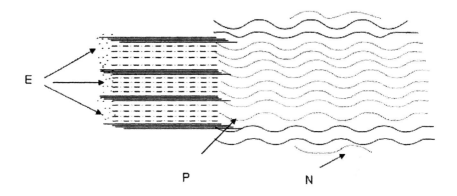

Figure 4. Microfibril structure (E: elementary fibrils; P: paracrystalline region; N: non-cellulose polysaccharides).

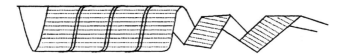

Figure 5. Helical structure of a microfibril.

Within a native cell wall, the successive cellulose chains in an elementary fibril as well as in a microfibril are joined together by hydrogen bonding. But apart from H-bonds, successive microfibrils may have an appreciable amount of non-cellulosic material in between. However, the structure is substantially altered by pulping and bleaching. After the cell wall is delignified and the hemicelluloses and pectin are extracted, there is only a little non-removable lignin between successive microfibrils. In the case of bleached kraft pulp, the microfibrils would be attached to each other through hydrogen bonding. The boundary between one microfibril and the next, which is typically well defined in native wood, would be minimal after chemical pulping and bleaching.

It may be noted that the definitions of microfibrils are several decades old, and the concept of nanosized structures was not well understood at that time. This explains the apparent anomaly that fibrils having the dimensions of 8-38 nm are called microfibrils rather than nanofibrils. In current terms, structures with significant dimensions less than 100 nm are generally considered to be nanoscale materials. This anomaly in nomenclature should be rectified while defining a "microfibre".

What Is a Fibre?

A fibre has been defined as a slender and greatly elongated solid substance. In plant cells, a fibre has been traditionally defined as an elongated sclerenchymatous cell with tapered ends. For structural applications in composites, a fibre should provide reinforcement; it should have high strength and stiffness that would impart superior mechanical properties to composites. For this purpose, it needs to have a minimum aspect ratio (length/diameter ratio) in order to allow sufficient stress transfer, and thereby act as an effective reinforcing agent. Below a certain aspect ratio, the material loses its reinforcing potential, and may be referred to as a "particle". In discussing composite materials, particles used for their bulk rather than mechanical properties are referred to as "fillers" rather than "fibres".

There are a couple of parameters related to the fibre length and aspect ratio that determine its performance in a composite. These are:

I. Stress transfer aspect ratio, and
II. Critical fibre length and critical fibre aspect ratio

The significance of these two parameters is discussed below. Both these theories are based on the assumption that there is perfect adhesion between the fibre and the matrix.

Stress transfer aspect ratio 1): The shear lag model originally proposed by Cox (*14*) is most widely used to describe the effect of loading on an aligned short-fibre composite, and is considered in this study. The stress fields acting along the length of the composite under an applied load σ_f are shown in Figure 6 for an elastic-perfectly plastic matrix, i.e., a matrix that is elastic up to the yield stress, and then strains indefinitely at the yield stress. The corresponding variations in fibre tensile stress along the length of a fibre are shown in Figure 6(e). The tensile stress is zero at the fibre ends, and a maximum at the centre. At the same time, the interfacial shear stress is zero at the centre, and a maximum at the fibre ends [Figure 6(f)]. A comparison between the Figures 6(b) and 6(d) shows there is a minimum fibre length for a given diameter, and consequently a minimum fibre aspect ratio, beyond which the fibre strain at the centre approaches the strain of the matrix. Above this aspect ratio, the fibre stress curve reaches a plateau region at the centre [Figure 6(e), Case A], with a corresponding interfacial shear stress of zero around the middle of the fibre [Figure 6(f), Case A]. This aspect ratio for which the peak (central) stress in the fibres closely approaches the maximum possible (at which the fibre strain at the centre approximates the value being imposed on the composite) is known as the *stress transfer aspect ratio,* denoted by s_t (*15*). If the aspect ratio of the same fibre is below this value, the matrix strain exceeds that of the fibre throughout the whole length of the fibre. In such cases [observed in Figure 6(e), Case B], the fibres do not provide efficient reinforcement, because they carry much less load than longer fibres in the same system. In other words, below s_t, the stress is not entirely transferred from the matrix to the fibre, and hence, the composite fails before the full reinforcing potential of the fibre is attained.

Therefore, an entity with a well-defined aspect ratio provides a consistent reinforcement in a given system if it has a minimum aspect ratio of s_t. This forms a basis for defining a fibre in terms of its reinforcing potential. However, the exact aspect ratio beyond which a material provides sufficient reinforcement and acts as a fibre, and below which it fails to attain its full reinforcing potential, differs widely depending on the matrix where it is used. From theoretical considerations, it has been shown (*15*) that the value of s_t is of the order of 30 for polymer matrix composites. The value will be greater with increasing stiffness of the fibre in a given polymer. More specifically, for glass fibres in a polyester resin matrix, a fibre length of 10 fibre diameters was shown to provide full reinforcement. An aspect ratio of 10 was also found to be sufficient for natural fibres in thermoplastic matrices (*16*). For most commercially interesting composites, however, a fibre provides little reinforcement below aspect ratio 5, and usually reaches its full reinforcing potential at an aspect ratio 50 and above. Therefore, for the purpose of defining a fibre, we may assume that a minimum aspect ratio of 50 is needed to achieve full reinforcement potential, i.e.,

$$L/d \geq 50 \tag{1}$$

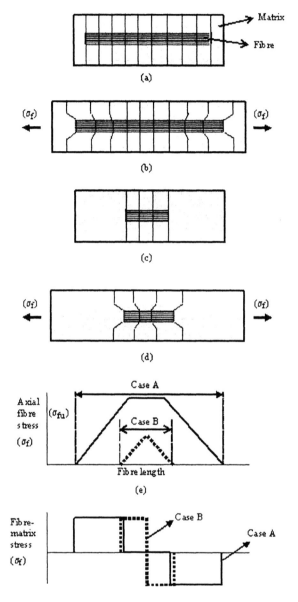

Figure 6. Single fibre composite element in an elastic-perfectly plastic matrix
(a) unstressed with s > s₁ (b) stressed under an axial load σf with s > s₁,
(c) unstressed with s < s₁ (d) stressed under an axial load σf with s < s₁,
(e) variation of tensile stress along the length of the fibre
(Case A: for s > s₁; Case B: for s < s₁),
(f) variation of fibre-matrix interfacial shear stress along the length of the fibre
(Case A: for s > s₁; Case B: for s < s₁).

where L = fibre length and d = fibre diameter.
As mentioned above, the exact value varies from one matrix to another.

Critical length II): As discussed above, according to the theory of reinforcement due to slip between a fibre and the matrix, in a representative volume element in a perfectly aligned fibre composite, the tensile stress (σ_c) is transferred from the matrix to the fibre through interfacial shear stress (τ) near the fibre ends. As the composite strain is increased, the yield or sliding spreads along the length of the fibre, raising the tensile stress in the fibre as the interfacial shear stresses increase. If the fibre is long enough, the tensile stress will eventually reach the strength of the fibre, and it will break at its midpoint. The minimum length needed for this is referred to as the *critical length.* The value of L_c may be derived by equating the force needed to break the fibre cross-section with the total force introduced by shear of the surrounding matrix. The maximum stress in the fibre occurs at the fibre midpoint and is obtained by a simple force balance:

$$\sigma_{fmax} \cdot 7\pi r^2 = \tau.2.\pi r(L_c/2) \tag{2}$$

where σ_{fmax} is maximum axial stress in the fibre at the midpoint, r is radius of the fibre, τ shear stress at the fibre-matrix interface and L_c half the length of the fibre, which is called the critical fibre length in this case. Below the critical length L_c, the fibres cannot be broken. To determine the value of L_c, the maximum fibre stress is set equal to fibre strength σ_{fu}, and eq (2) may be simplified to give

$$\sigma_{fu} = \tau.L_c/r \tag{3}$$

If a fibre has a length $L \geq L_c$, and is embedded to a depth of $L_c/2$ on either side of a propagating crack, then it would break rather than pull out, as depicted in Figure 7. On the other hand, if $L < L_c$, or if one end is embedded to a depth of less than $L_c/2$, the fibre will be pulled out of the matrix without fracture (*17*). Accordingly, for a given matrix and an associated fibre-matrix interface, the *critical fibre aspect ratio* (s_c) is defined as

$$s_c = L_c/d = \sigma_{fu}/2\tau \tag{4}$$

which is deduced by substituting the value of L_c from eq 3.
An example of the critical aspect ratio for a cellulosic material may be deduced from the work done by Pervaiz (*18*). In this study, hemp fibres of 65

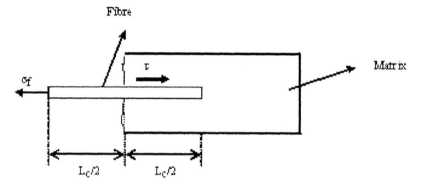

Figure 7. Forces acting on a single fibre partially embedded in a matrix when subjected to a tensile stress σ_f (τ = interfacial shear stress; L_c = critical fibre length).

μm diameter gave a critical length of 3.4 mm in polypropylene. Therefore, s_c for hemp had a value of 52 in polypropylene. The value of s_c for a given system can be determined experimentally by noting the minimum length at which a fibre half-embedded in a matrix fractures at the maximum load, instead of pulling out of the matrix. However, as in the case of s_t, s_c also depends on the interaction of the fibre with the matrix, and is not an intrinsic property of the fibre itself. Hence, s_c for the same fibre also varies with matrix in which it is used.

Therefore, both the parameters that characterize a fibre in terms of its performance in a composite depend on the properties of the matrix as well. Hence, it is not possible to define an entity as a "fibre" based solely on its intrinsic properties.

Deriving the Definition of a "Microfibre" as a Fibre

In accordance to earlier discussions, the proposed definition for microfibres as fibres for structural applications is based on the following basic principles:

I) Microfibrils, as have been traditionally defined, do not have any distinct interface separating each other once the cell wall is delignified and the hemicellulose and pectin are extracted. Traces of lignin may remain in between the microfibrils even in bleached kraft pulp after thorough delignification. However, the microfibrils also have amorphous regions in them, which make the intrafibrillar structure weak. Therefore, in terms of crystallinity, there is no distinct interface beyond an elementary fibril that makes one microfibril structurally defined, particularly when the cell wall structure is thoroughly delignified.

II) Unlike a microfibril, a microfibre is to be considered a fibre for structural applications. Therefore, the definition of a microfibre naturally has to conform to the definition of a fibre.

III) For a consistent definition, a microfibre should be defined as having a diameter in the range of microns, as opposed to nanometers as in the case of microfibrils.

Eq 1 and eq 4 present two different perspectives for defining a fibre based on its performance in a composite. Of these two, s_t indicates the minimum fibre aspect ratio needed to attain the full reinforcing potential of the fibre in a given matrix. It is s_t that controls the stiffness of a composite. On the other hand, s_c determines whether a fibre will pullout or fracture when it reaches its ultimate stress, and is critical in determining the fracture toughness of a composite (*17*).

Hence, the definition of a microfibre is based in this study on the value of s_t. However, there is no way to define quantitatively a universal value of s_t, and hence a microfibre, without knowing the properties of both the fibre and the specific matrix. Instead, a value of 50 for s_t is assumed, ensuring sufficient margin of error over the experimental value of 10 obtained by Facca et al. *(16)* for natural fibres in a thermoplastic matrix.

With all the above considerations, for structural materials *a microfibre is hereby defined as a fibre consisting of continuous cellulose chains with negligible lignin and hemicellulose content, and having a diameter between 0.1 to 1 µm, with a minimum corresponding length of 5 to 50 µm.* In the longitudinal direction, it consists of alternating crystalline and amorphous zones of cellulose. Laterally, however, the chains are linked together by a combination of hydrogen bonding, amorphous cellulose molecules, and traces of lignin and hemicellulose not removed through the pulping processes.

Table I summarizes the diameter and aspect ratio of cellulose microfibres in comparison to wood fibres, microfibrils, and cellulose whiskers derived from wood. Softwood fibres having a length in the range of 2 mm are considered in this case. It should be noted that the values of the diameter and aspect ratio may vary widely with the specific source of cellulose. For example, cellulose whiskers derived from tunicin typically have an aspect ratio several times greater than those from wood.

Typical SEM images of fibres, microfibrils, and microfibre from wood are shown in Figure 8.

Table I. Diameter and aspect ratio of fibres, microfibrils, microfibres, and whiskers from wood

	Diameter (µm)	Aspect ratio range
Fibre	5 – 25	100 – 400
Microfibre	0.1– 1	> 50
Microfibril	$5 \times 10^{-3} – 10 \times 10^{-3}$	Undefined
Whisker	$\sim 3 \times 10^{-3}$	$20 – 70^1$

The lower value is based on a crystalline length of 60 nm *(4)* obtained by X-ray studies and microscopic imaging, and the higher value is based on degree of polymerization measurements, giving average whisker length as 180 – 200 nm *(19)*.

Microfibre Generation and Its Reinforcing Potential

In our laboratory, cellulose microfibres of diameters less than 1 µm and with aspect ratio greater than 50 have been generated and isolated from bleached kraft softwood pulp fibres. This was achieved through a combination of high

183

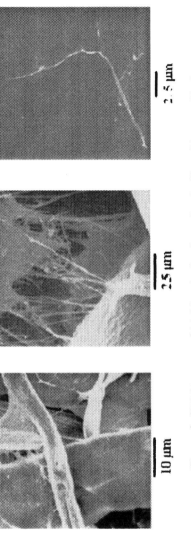

Figure 8. SEM images of (a) bleached softwood pulp fibre (b) microfibrils stretching out from the cell wall, and (c) a single microfibre

shear refining and crushing under liquid nitrogen, and subsequently passing the resulting fibres through a mesh size of 60. The details of the process have been described elsewhere (*20*).

The reinforcing ability of the microfibres was demonstrated by dispersing them in an aqueous solution of polyvinyl alcohol (PVA), and preparing films by solution casting (*21*). Table II shows a comparison of pure PVA films with composite films prepared with 5% loadings of bleached kraft pulp (BKP), microfibres (MF), and microcrystalline cellulose (MCC). The sample reinforced with 5% microfibres showed much higher tensile strength and modulus compared to the other samples. This illustrated that the microfibres defined above effectively act as fibres for structural applications, and the aspect ratio is high enough to provide sufficient reinforcement to the polymers in which they are dispersed.

Table II. Tensile properties of different cellulose reinforcements in PVA

	Peak strength (MPa)	*Stiffness (GPa)*
Pure PVA	43 ± 5	2.3 ± 3
PVA + 5% BKP	58 ± 8	3.9 ± 1
PVA + 5% MF	102 ± 2	5.4 ± 3
PVA + 5% MCC	72 ± 7	3.4 ± 4

It is challenging to draw a fair comparison between these values obtained in our laboratory, and studies done by other researchers. This is mainly because the use of microfibres as reinforcing agents has been scarce. Nevertheless, a comparison between mechanical properties of microfibrillated cellulose (MFC) and kraft pulp fibres would be relevant (*22*). MFC refers to cellulose fibres split and unraveled to expose smaller fibrils and microfibrils, the latter having diameters in the range 10–100 nm. Therefore, unlike MF, MFC forms an entangled mass, and does not have a distinct aspect ratio. In this study, when PF was used as a binder, and the specimens were compressed at 100 MPa, the MFC based composite showed a bending strength of up to 370 MPa, compared to about 220 MPa for the pulp fibre-based composite. The bending strength of the MFC-based composites peaked at as low as 10% PF content, i.e., 90% cellulose content. The stiffness, however, was very similar, around 18 – 19 MPa, for both the MFC-based and pulp-based composites. A comparison with Table 2 with only 5% cellulose reinforcement indicates that MF acts as a stronger reinforcing agent than MFC.

185

Conclusions

This paper analyzed the configuration of cellulose chains starting with the structure of a glucose molecule. The arrangement of cellulose crystallites to form elementary fibrils through hydrogen bonding, and subsequently the formation of microfibrils with these elementary fibrils was discussed in detail. Subsequently, the need to define "microfibres" with a minimum aspect ratio was highlighted, as opposed to microfibrils which are attached to the cell wall at one end. The significance of stress transfer aspect ratio and critical fibre aspect ratio in determining the performance of a fibre in a matrix was discussed. With the consideration that a minimum aspect ratio of 50 ensures that the strain at the centre of the fibre approaches the matrix strain, it was proposed that microfibres should have a minimum aspect ratio of 50 for adequate stress transfer from the matrix to the fibre. With this consideration, a cellulose microfibre was defined as a fibre consisting of continuous cellulose chains with negligible lignin and hemicellulose content, and having a diameter between 0.1 to 1 μm, with a minimum corresponding length of 5 to 50 μm. These microfibres were generated in the laboratory, and were shown to provide substantial reinforcement in PVA films produced by film casting.

References

1. Mühlethaler, K. *Biochim. Biophys. Acta* **1949**, *3*, 15.
2. *Plant cells;* Clowes, F.A.L.; Juniper, B.E., Eds.; Blackwell Scientific Publications, 1968.
3. Rydholm, S.A. *Pulping Processes;* Interscience Publishers, 1965, pp. 11.
4. Ohad, I.; Mejzler, D. *J. Polym. Sci.: Part A* **1965**, *3*, 399.
5. Frey-Wyssling, A; Mühlethaler, K. *Makromol. Chem.* **1963**, *62*, 25.
6. *Cellular ultrastructure of woody plants,* Côté W. A. Jr., Ed,; Advanced Science Seminar, Pinebrook Conference Center, 1964.
7. Preston, R.D. *Faraday Soc. Disc.* **1951**, *11*, 165.
8. Günther, I. *J. Ultrastruct. Res.* **1960**, *4*, 302.
9. Hodge, A.J.; Wardrop, A.B. *Nature* **1950**, *165*, 2725.
10. *Vogel, A. Makromol. Chem.* **1953**, *11*, 111.
11. Fengel, D.; Wegener, G. *Wood: chemistry, ultrastructure, reactions;* W. de Gruynter, 1984.
12. Preston, R.D.; Kuyper, B. *J. Exp. Bot.* **1951**, *2*, 247.
13. Manley, R. St. J. *Nature* **1964**, *204*, 1155.
14. Cox, H.L. *Br. J. Appl. Phys.* **1952**, *3*, 72.
15. Hull, D. *An introduction to composite materials,* 2nd ed., Cambridge University Press, Cambridge, Great Britain, 1996, pp. 105-121.

16. Facca, A.G.; Kortschot, M.T.; Yan, N. *Journal of Composites A,* submitted on March 16, 2005.
17. Piggott, M. *Load bearing fibre composites,* 2nd ed., Kluwer Academic Publisher, Norwell, MA, USA, 2002, pp. 156 – 161.
18. Pervaiz, M. Design and process optimization of reinforced natural fibre mat (NMT) composites. *Thesis on Master of Science in Forestry,* 2003, Faculty of Forestry, University of Toronto.
19. de Souza Lima, M.M.; Borsali, R. *Macromol. Rapid* Commun., **2004**, *25,* 771.
20. Chakraborty, A.; Sain, M.; Kortschot, M. *Holzforschung,* **2005**, *59,* 102.
21. Chakraborty, A.; Sain, M.; Kortschot, M. *Holzforschung,* accepted on 23 September, 2005.
22. Nakagaito, A.N.; Yano, H. *Appl. Phys. A,* **2004**, *80,* 155.

Chapter 13

Dispersion of Soybean Stock-Based Nanofiber in Plastic Matrix

B. Wang and M. Sain

Centre for Biocomposites and Biomaterials Processing, Faculty of Forestry and Department of Chemical Engineering and Applied Chemistry, Earth and Science Centre, University of Toronto, 33 Willcocks Street, Toronto, Ontario M5S 3B3, Canada

Plant stem fibers primarily contain cellulose, hemicellulose, pectin, and lignin. These are bundles of cellulose nanofibers with a diameter ranging between 10 to 70 nm and lengths of thousands of nanometers. The mechanical performance of the cellulose nanofibers is comparable to other engineering materials such as glass fibers, carbon fibers etc. In this project, the cellulose nanofibers were extracted from soybean stock by chemo-mechanical treatments. Since they are new types of reinforced material used, the composition, dispersion and morphological properties of the nanofiber were investigated and their properties compared with those of hemp nanofibers. The matrix polymers used in this project were polyvinyl alcohol (PVA) and polyethylene (PE). These nanofibers were characterized by atomic force microscopy (AFM) and transmission electron microscopy (TEM). X-ray diffraction (XRD) results showed the estimated crystallinity of soybean stock nanofibers. One of the major challenges faced was the incompatibility of the nanofibers and PE. To synthesize a biocomposite with PE, a number of mixing principles were explored. Improved dispersion of nanofibers was achieved by adding ethylene-acrylic oligomer emulsion as a dispersant. Selective chemical treatments increased cellulose content of soybean stock nanofibers from 41% to 61%. Nanofibers reinforced PVA films demonstrated at least a 4-5-fold increase in tensile strength, as compared to the untreated fiber/PVA film. In solid phase nanocomposites, improved mechanical properties were achieved with coated nanofibers.

Introduction

In the plant stem, the fiber cell wall consists not only of cellulose, but also of hemicellulose, pectin and lignin. The properties of each constituent contribute to the overall properties of the fiber. Figure 1 illustrates the hierarchical microstructure of cellulose fiber bundles. The plant possesses cell walls where cellulose nanofibers are embedded in a gel-like matrix (*1*). The cellulose molecules are always biosynthesized in the form of nanosized fibrils; up to 100 glucan chains are grouped together to form elementary fibrils, which aggregate together to form cellulose nano-sized microfibrils or nanofibers (*2*). They have diameters in the range of 5-50 nm and lengths of thousands of nanometers (*3, 4, 5*). These nanofibers provide strength to the plant stem fiber. The mechanical performance of cellulose nanofibers in terms of the tensile strength and Young's modulus is comparable to other engineering materials such as glass fiber, carbon fiber, etc. Therefore, the cellulose nanofibers can be considered to be an important structural element of natural cellulose in engineering applications.

Many studies have been done on extracting cellulose nanofibers from various sources and on using them as filler in composite manufacturing (*6, 7*). These microfibrils can be extracted from the cell walls by two types of isolation processes: purely mechanical methods or a combination of both chemical and mechanical methods. A purely mechanical process can produce refined, fine fibrils several micrometers long and between 50 to 1000 nm in diameter. However, these processes result in degradation of the cellulose. In contrast, chemo-mechanical treatments can extract cellulose nanofibers from the primary and secondary cell walls without degrading the cellulose. A chemo-mechanical process can also achieve finer fibrils of cellulose (cellulose nanofibers), ranging between 5 and 50 nm diameter.

The fairly new idea of bionanocomposites, in which the reinforcing material has nanometer dimensions, is emerging to create the next generation of novel eco-friendly materials with superior performance and extensive applications in medicine, coatings, packaging, automotive and aerospace applications etc. Biodegradable nanocomposites which have superior thermal, barrier and mechanical properties compared to today's biomaterials can be synthesized from biopolymer and nano-sized reinforcements.

Raw materials for these new nano-structured composites are usually natural biopolymers (PLA, PVA and cellulose esters). In the present paper, hemp and soybean stock nanofibers were used as reinforcements for the manufacture of cost effective and environmentally-friendly composites. Cellulose nanofibers have not been used extensively in the common thermoplastics (e.g. polyethylene and polypropylene), as they are more expensive than wood flour and not readily available. In addition, they face the same incompatibility problem inherent in wood-plastic composites: the cellulose tends to agglomerate and the resulting composite is more susceptible to moisture than the neat plastic.

Crystalline cellulose chains connect to each other by hydrogen bonding which results in the agglomeration or entanglement of the nanofibers in polymer

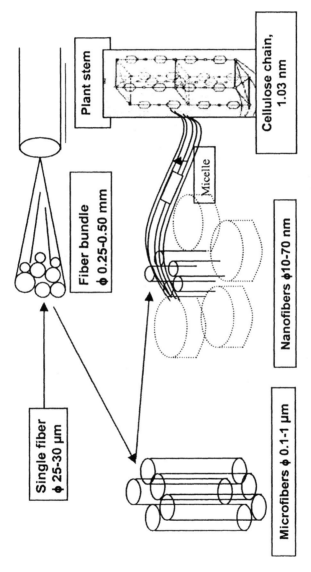

*Figure 1. Hierarchal structure of cellulose to fiber bundle (adopted from:
http://www. botany.utoronto.ca/courses/new_BOT251Y1Y/TSage/TS_lec_4.pdf.*

matrix. High energy is required to overcome this strong hydrogen bonding. In order to reduce the interaction between hydroxyl groups, the nanofibers obtained after chemo-mechanical treatments are kept in water suspensions. Water is the most widely used carrier to disperse cellulose nanofibers. Therefore the use of nanofibers has been mostly restricted to water soluble polymers, which are incorporated into a cellulose aqueous suspension. But to expand the horizon of bio-based nanocomposites for high-end applications, it is necessary to reduce the entanglement of the fibrils and improve their dispersion in the polymer by surface modification of fibers, without deteriorating their reinforcing capability. Attempts have been made to modify the nanofiber surface in order to improve their dispersion in non-polar solvents (8, 9, 10).

In the present paper, the processing of nanocomposites was studied with the aim of achieving a deeper understanding of compatibility between different phases and dispersion of nanoreinforcements in the biopolymer matrix. Moreover, the development and optimization of the existent and new manufacturing techniques was established. Characterizations methods such as TEM and AFM, important techniques in determining nanoscale properties, are usually used for metallic materials and one challenge in this project is to use these techniques for biomaterials. Processing, characterization, and functional properties of this new class of nanocomposites can be further optimized.

The chemo-mechanical method has been used to produce cellulose fibrils, essentially within the diameter range of 50-100 nm and hence these fibrils have been referred to as cellulose nanofibers. The goal of this research was to explore the dispersion of nanofibers from soybean stock in PVA and PE and compare their properties with hemp nanofibers. The use of ethylene-acrylic oligomer aqueous emulsion can be effective in improving the compatibility between non-polar plastics such as PE, PP and polar additives such as pigments, chalk, cellulose, etc. Further research work is required for an improved understanding of dispersion mechanisms of nanofibers in the solid polymer matrix.

Materials and Methods

Materials

The raw materials used in this study were industrial grade hemp fiber supplied by Hempline Inc., Canada and soybean stock from Ontario Soybean Producers. Reagent grade chemicals are used for fiber extraction, namely, sodium hydroxide (NaOH) and hydrochloric acid (HCl). Polyvinyl alcohol (Dupont) is a water-soluble polymer. Ethylene-acrylic oligomer aqueous emulsion was used as a dispersant. Copolymers[1] of these substances contain functional groups which influence the solubility of the material. Acrylic oligomer emulsions are easy to handle because of their aqueous phase. They follow the general trend towards solvent-free formulations as an innovative application in improving surface properties. The emulsion has a solids content of 35 %, and it has a very narrow distribution of particle size in the 100 - 150 nm range.

Extraction of Nanofiber

The isolation of nanofibers is a multi-step process (Figure 2). Chemical and mechanical treatments are applied onto the fiber to make nanofibers. The chemical treatment includes pre-treatment, acid hydrolysis and alkaline treatment. The mechanical treatment includes cryocrushing by liquid nitrogen and high-pressure defibrillation (*12, 13*). Defibrillation is a process of further opening up of single fibers by applying high shear (*8*).

Generally, the first step for all of the fiber surface treatments is the mercerization process (pre-treatment process) which causes change in the crystal structure of cellulose. Cellulose fibers were soaked in sodium hydroxide solution at room temperature overnight to swell the cell wall, enabling chemical molecules to penetrate through the crystalline region of the cellulose. This removes the lipid membranes and improves the efficiency for later chemical treatments (*5*).

The chemical treatment removes the starch, hemi-cellulose, and lignin/pectin surrounding the cellulose and destroys the bonding between fiber bundles. The alkaline treatment also improves the adhesion between fiber and polymer and breaks the fiber into smaller fibers. Acid hydrolysis with hydrochloric acid followed by alkaline treatment with mild sodium hydroxide is applied to further remove lignin, hemicellulose and pectin. The fiber obtained after the chemical treatment contains mainly alpha-cellulose with some hemi-cellulose and lignin.

The mechanical treatment was comprised of two parts: power crush with liquid nitrogen and high pressure defibrillation. These processes reduce the size of the fibers to form cellulose nanofibers and improve the surface of the fibers for better adhesion between the fiber and the polymer matrix in the composite making stage. The sample was first cryocrushed with liquid nitrogen to reduce the fiber length. The objective of the cryocrushing is to form ice crystals within the cells. When high impact is applied on the frozen fiber, ice crystals exert pressure on the cell wall, causing it to rupture, thereby liberating the microfibrils. The sample was then passed into the defibrillator to crush the cell wall and fully release the nanofiber. In this way, nanofiber was produced. The suspension was propelled through valves under the influence of a high-pressure gradient. The high-pressure defibrillation process led to the individualization of cellulose nanofibers. Finally, the suspension obtained from the defibrillator was dispersed in a lab scale disintegrator.

192

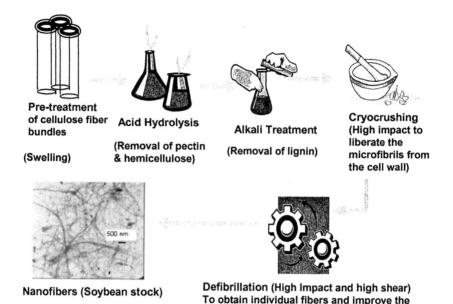

Pre-treatment of cellulose fiber bundles

(Swelling)

Acid Hydrolysis

(Removal of pectin & hemicellulose)

Alkali Treatment

(Removal of lignin)

Cryocrushing (High impact to liberate the microfibrils from the cell wall)

500 nm

Nanofibers (Soybean stock)

Defibrillation (High Impact and high shear) To obtain individual fibers and improve the dispersion of fibers

Figure 2. Isolation of nanofibers.

Preparation of Composites Film

This project focused on synthesizing the nano-biocomposite using polyethylene (PE) in the solid phase and poly (vinyl alcohol) (PVA) in an aqueous phase as the polymer matrix.

To synthesize the biocomposite, solid PVA was mixed with a water suspension of nanofiber. The mixture was then heated and mixed at 80-90°C for 15 to 30 min. After the solution became viscous, the mixture was poured onto a plastic plate and then heated in a 60°C oven for overnight until it was completely dry. The films were finally removed (by peeling) from the trays and placed in desiccators to avoid moisture exchange.

A solid phase compounding method was used to mix the nanofiber with PE. 5% by weight of the coated nanofibers were added to molten PE in a laboratory kneader. The composite was mixed at the melting temperature of PE. After the sample was well-mixed, it was pressed at high temperature with WABASH Hot Press into sheet form. The challenge in nanofiber and PE compounding is to make the nanofiber and matrix compatible to achieve high mechanical performance. Nanofiber is hydrophilic whereas polyethylene is hydrophobic in nature. This issue was addressed by coating the nanofiber with an acrylic aqueous emulsion, which acts as a bridge between the hydrophilic and hydrophobic components.

Characterization Methods

Chemical Characterization of Fibers

Fiber samples of all the raw materials contain holocellulose, pectin and lignin. An acid-chlorite solution was used to dissolve the lignin component around the cellulose strands of the fibers, leaving behind only holocellulose (hemicellulose and alpha-cellulose).

Holocellulose determination: For each sample, 0.70±0.05 g of Wiley milled fiber was placed in a 125 ml Erlenmeyer flask. 10 ml of acetic acid-sodium hydroxide was added followed by 1 ml of sodium chlorite solution. The flasks were swirled until the reagents were well mixed with the fibers. The flasks were placed in a water bath maintained at 70 °C. After each of 45 minutes, 90 minutes, 150 minutes and 220 minutes, an additional 1 ml of NaClO$_2$ was added. After 4 hrs holocellulose was filtered through coarse crucibles and washed with 100 ml of 1 % acetic acid and with 10 ml of acetone. The crucibles were left in a conditioning chamber for four days. On the fourth day, holocellulose was measured and calculated in percentage based on oven-dried weight.

α Cellulose determination: After chlorite reduction, the sample contains only three basic cellulose polymers. At this point a further acid reduction is carried out to remove all other chemical components except pure cellulose. Holocellulose is the sum of hemicellulose and alpha cellulose. 3 ml of 17.5% NaOH solution was added to holocellulose in each crucible. After 5 minutes, another 3 ml of NaOH was added. After 35 minutes from the second addition, 6 ml of water was added to each crucible. Then each crucible was dried under a mild suction and then 60 ml of distilled water was added, followed by an addition of 5 ml of 10% acetic acid. The crucibles were then washed with two 10 ml additions of acetone and dried in an oven at 103 °C. The mass of each crucible was determined after overnight drying. Based on OD weight of the fibers, alpha-cellulose was reported.

Hemicellulose determination: Subtracting the amount of alpha-cellulose from total holocellulose gives the combined hemicellulose and beta and gamma cellulose content.

Klason lignin and soluble lignin determination: Lignin content was determined based on TAPPI T 222 om-02.2002 (acid-insoluble lignin in wood and pulp) and TAPPI Useful Method UM250 (raw material and pulp-determination of acid-soluble lignin). Sulfuric acid was used to hydrolyze the polysaccharides in the fibers, leaving behind the lignin. 1 g of sample was placed in a 50 ml beaker and 15 ml of 72% sulphuric acid were added with continuous stirring to each beaker. The beaker was placed on a water bath maintained at 18-20 °C. The beakers were placed there for 2 hrs with frequent stirring. Then the samples were transferred into 1L flasks, to each of which, 560 ml of distilled water was added, so as to dilute the concentration of sulphuric acid to 3%. The flasks were placed on a hot plate and boiled for 4 hours maintaining the fluid level in each flask at 560 ml. For another 4 hrs, insoluble lignin was allowed to settle down. The mixture was transferred to a medium crucible and filtered. The filtrate was kept for the determination of the soluble lignin. The mixture was washed with at least 500 ml of distilled water until it became neutral. The crucibles were then kept in an oven overnight at 103 °C and weighed the next day to determine the insoluble lignin content as a percentage of extractive free oven dry samples.

Total Lignin (TL) % Insoluble Lignin (LO) % + Soluble Lignin (SL) %

The TAPPI method UM250-85 was used to determine the soluble lignin content of the fiber sample, a small quantity of filtrate from the above stage was poured in a cuvette and the absorbance was measured at 205nm wavelengths using a UV Spectrophotometer, with 3% sulphuric acid solution as a blank. Soluble lignin content was determined as:

SL= [(Absorbance /110) x 0.575]/ (oven dry mass of fiber sample) x 100

Cellulose Nanofiber Morphological Characterization

Transmission electron microscopy (TEM) observations were made using with a Philips CM201 operated at 80 kV. A drop of a dilute cellulose nanofiber suspension was deposited on carbon-coated grids and allowed to dry. AFM images were obtained using a Digital Instruments Dimension 3100 AFM with a nanoscope IIIa controller. The system was operated in tapping mode with DI tapping mode tips having a resonant frequency of 280 kHz.

Crystallinity Determination by Powder X-ray Diffraction Method (PXRD)

X-ray crystallography is carried out to better understand the structure of the native cellulose of untreated fibers, regenerated celluloses after chemical purification of the fibers and lastly of cellulose nanofibers. The aim of X-ray diffraction is the evaluation of scattering direction and intensity diffracted by atom planes, according to Bragg's law. To estimate the crystallinity of the sample, the total area under the curve is calculated and then the amorphous portion is subtracted from the total area to give only the crystalline portion. A D8 Advance Bruker AXS diffractometer, Cu point focus source, Graphite monochromator, and 2D-area detector GADDS system were used. Samples were analyzed in transmission mode. Percentage crystallinity was determined after various stages of the chemo-mechanical treatment and of nanofibers.

Tensile Strength Test

The mechanical behavior of nanofiber-PVA film and nanofiber-PE nanocomposite was tested by a Sintech-1 machine model 3397-36 in tensile mode with a load cell of 50 lb followed by ASTM D 638. The specimens were cut in a dumbbell shape with a die ASTM D 638 (type V). Tensile tests were performed at a crosshead speed of 5 mm/min. The values reported in this work result from the average of at least five measurements.

Results and Discussions

Chemical Characterization of Fibers

It is desirable to retain pure cellulose, whose crystalline form and high packing density result in a stronger composite, while removing components such as hemicellulose and lignin, which are amorphous, can easily absorb chemicals, and tend to reduce the mechanical strength. After cellulose nanofibers are embedded in the pectin/hemicellulose matrix, acid and alkali treatments are applied. These lead to almost pure cellulose fibers, removing not only the lignin, hemicellulose and pectin, but also the surface impurities, waxy substances and hydrophilic hydroxyl groups. Figures 3-6 confirm removal of a significant portion of lignin, hemicellulose and pectin.

At the end of the chemical treatment for a soybean stock, the α-cellulose content increased from 41% to 61%, as shown in Figure 3. A similar trend was seen in hemp fiber (Figure 4), where the a-cellulose content increased by 20% and the hemicellulose content reduced to 1.6%, with only 3-4% lignin left compared to 6% initially (Figure 6). The high lignin content of soybean stock significantly decreased from 16% to 4% (Figure 5). After chemical treatment, it was found that hemp fibers contained both soluble and insoluble lignin, while soybean stock fibers contained only insoluble lignin. Chemical analysis of these fibers after each stage of purification showed an increase in cellulose content and decreases the lignin content.

Dispersion of Cellulose Nanofibers

Cellulose fibrils have a high density of -OH groups on the surface, which have a tendency to form hydrogen bonds with adjacent fibrils, reducing interaction with the surrounding matrix. Agglomeration is the formation of groups of cellulose fibers due to the hydrogen bonds between each of them (Figure 7). This formation of hydrogen bonds accounts for the hydrophilic

196

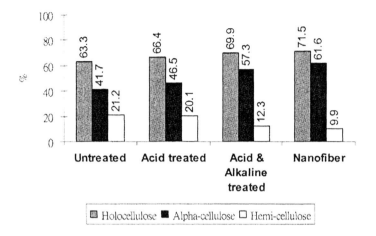

Figure 3. Chemical changes of soybean stock.

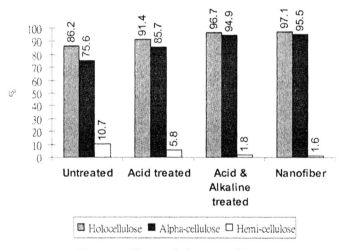

Figure 4. Chemical changes of hemp.

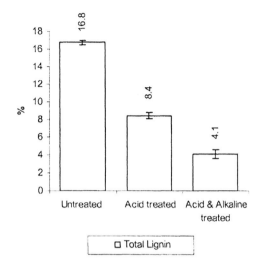

Figure 5. Lignin content changes of soybean stock.

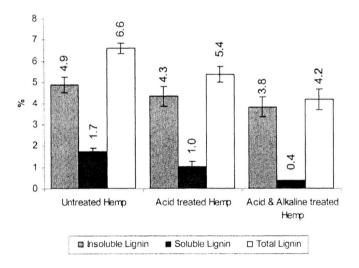

Figure 6. Lignin content changes of hemp.

198

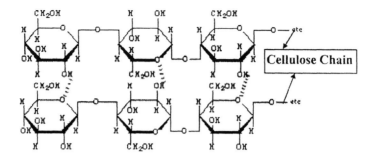

Figure 7. The hydrogen bonds between the cellulose chains.

properties of the cellulose fibers. As a result of the tendency for cellulose fibers to group together, it is problematic to combine the hydrophilic nanofiber with a non-polar polymer matrix. The cellulose fibers are not compatible with the hydrophobic polymer matrix and agglomerate, deteriorating their reinforcing capability. Overcoming strong hydrogen bonds requires high energy. The high pressure and high energy were imparted to the cellulose fibers to defibrillate fibrils intensively and improve the fiber dispersion.

It is possible to reduce the entanglement of the fibrils and improve their dispersion in the polymer by fiber surface coating. In this project, ethylene-acrylic oligomer was used as a dispersant. A suggested scheme for this process can be represented as shown in Figure 8. The cellulose chains may associate with ethylene-acrylic oligomer in two different ways. In case I, where the amount of oligomer used is limited, the acrylic oligomer may only partially disperse the cellulose fibers by forming chemical linkages with hydroxyl groups of cellulose and ester groups of the oligomer. In that case the dispersion would be incomplete; however in case II where oligomer concentration is high enough, the acrylic oligomer could block all effective hydroxyl groups of cellulose molecules by forming chemical bonds and repel both sides of the cellulose chains to avoid agglomeration of nanofibers. This suggested mechanism can provide the best dispersion effect on the nanocomposite.

TEM was used to investigate the size of the dispersed nanofibers. TEM can work on the wet surface based materials. TEM images (Figure 9) show that the nanofiber suspension with acrylic oligomer emulsion has better dispersion than the suspension without acrylic oligomer emulsion. This is because the acrylic oligomer reduces interaction between hydroxyl groups by hydrophobization. Nanofibers in their original form are rarely dispersed well in a plastic. The use of the acrylic oligomer made the nanofiber easier to disperse, enabling it to be dispersed more homogeneously in polymer matrix. The nanofibers are stabilized in aqueous suspension by acrylic groups on their surface.

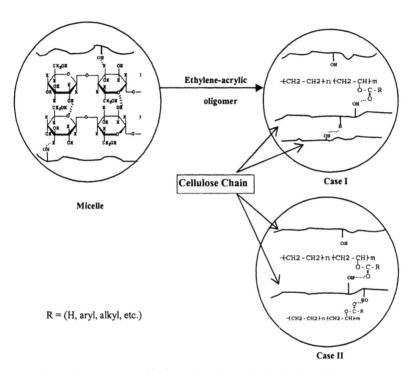

Figure 8. A suggested chemical scheme for the hydrophobization.

200

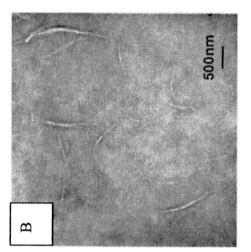

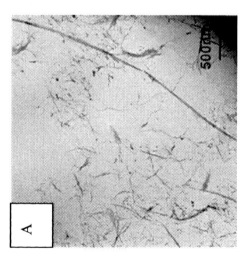

Figure 9. Transmission electron micrograph of cellulose nanofibers A) soybean stock, B) soybean stock coated with ethylene-acrylic oligomer emulsion.

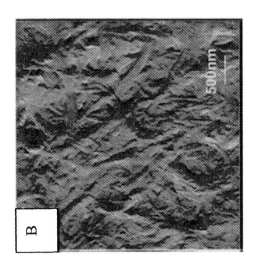

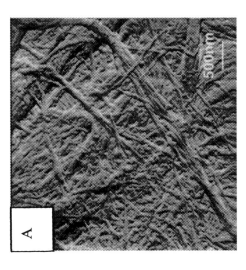

Figure 10. Atomic force micrographs of cellulose nanofibers A) soybean stock, B) hemp.

The suspension obtained after the high-pressure defibrillator was analyzed to determine diameters using AFM. Atomic Force Microscopy was used to investigate the surface properties and capture the surface images of the cellulose nanofibers. The atomic force micrographs (Figure 10) show the surface of air-dried soybean stock and hemp nanofiber. It is seen that the fibers are indeed nano-sized and the diameter of nanofibers is within the range of 50-100 nm. The network of nanofibers can also be seen in Figure 10.

Structural Characterization X-Ray Diffraction

According to X-ray testing on cellulose, cellulose is not made up of single perfect crystals. There are small crystalline structures called crystallites or nanocrystals, also known as micelles. They have length of about 600 Å and width ranged from 50 Å to 100 Å. Each are perfectly aligned cellulose chains. It is believed that these crystallites are connected to each other by disoriented amorphous zones (11). A highly crystalline region in the nanofibers provides high stiffness to the fibril and thus makes it less strong than less crystalline nanofibers. Therefore depending upon the application, high amorphousness may be favorable as it leads to increased ductility.

X-ray crystallography is used to investigate the crystallinity of the sample at different stages. Figure 11 is the pictorial view of the actual x-ray pattern of nanofibers. A sharp bright pattern represents the crystalline region of the nanofiber sample from soybean stock and hemp. In Figure 11 A), soybean nanofibers show a diffused and very dull pattern, indicative of very low crystallinity of the soybean nanofibers. The soybean stock nanofiber, even after chemical treatments, possesses 16% lignin and hemicellulose combined. These probably hinder the complete separation of nanofibers and also add to the amorphousness of the fibrils. The x-ray powder diffraction pattern is indicative of a crystalline material and shows peaks at $2\theta = 20.3°$ and $22.2°$. The peaks in the hemp pattern, shown in Figure 11 B), are sharper than those of the soybean stock, indicating higher crystallinity. Hemp had very little hemicellulose and lignin left insoluble after successive chemical treatments, leading to the more pure cellulose content and higher crystallinity. The peaks observed at $2\theta = 20.0°$ and $22.1°$ are highly prominent with low background indicating a crystalline material. It was observed that the soybean nanofiber had an estimated percentage crystallinity of about 48% and hemp nanofiber had a crystallinity of about 58%. As shown in Figure 12, the crystallinity of the samples increases after each stage of chemical and mechanical treatments because the amorphous regions are susceptible to water or chemical penetration. During the chemical treatment, the cellulose chain breakage would occur at the amorphous regions first. The amorphous regions in the cellulose chains are more vulnerable to the chemical hydrolysis and degrade before the crystalline region. As a result, the structure has an increasing crystallinity throughout the chemical process. During the mechanical treatment, the amorphous region is less stiff when compared to

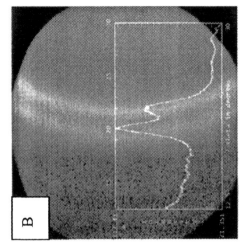

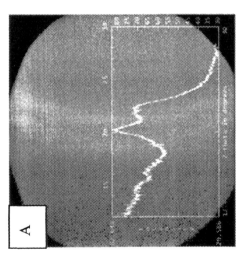

Figure 11. X-ray pattern to demonstrate the crystallinity of nanofibers A)soybean stock, B) hemp.

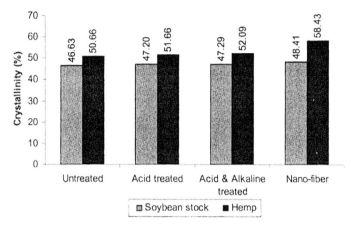

Figure 12. Crystallinity estimation after each stage of chemo-mechanical treatment for soybean stock and hemp fiber.

the crystalline region and is more susceptible to shear and stain under high pressure. Therefore, fiber disintegrates mostly at the amorphous region and the crystallinity of the final product increases.

Mechanical Performance of Nanocomposite

Figure 13 shows that the tensile strength at yield of a 10% nanoflber/PVA film increased by five-fold from that of the untreated flber/PVA film. Also it is observed that there is a two-fold increase in tensile strength from that of the pure PVA film. With a-5% soybean nanofiber reinforcement, the peak stress of the composite was approximately 102 MPa, compared to 129 MPa for a 10% nanofiber reinforcement. This shows the high reinforcing potential of the soybean nanofibers, and hence, they can be considered as an important structural element in engineering applications. Mechanical properties can be further enhanced if nanofibers can be dispersed more uniformly in the polymer matrix.

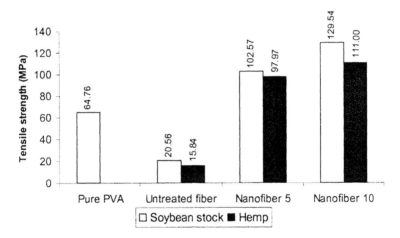

Figure 13. Tensile strength at yield of nanocomposites and pure polymeric matrix.

For solid phase nanocomposites, it was found that coating the nanofibers introduced into the molten PE phase improved their dispersion. Optical microscopy (Figure 14) compares the solid phase nanocomposite with and without using the acrylic oligomer during processing. In Figure 14 A), which corresponds to not using acrylic oligomer, many agglomerated fibers of micrometer size are visible. In contrast, in Figure 14 B), it is observed that there is a very limited amount of microfibers, the fiber diameters being much smaller. This demonstrates that the nanofibers were dispersed more effectively in the composite when they were coated with the acrylic oligomer.

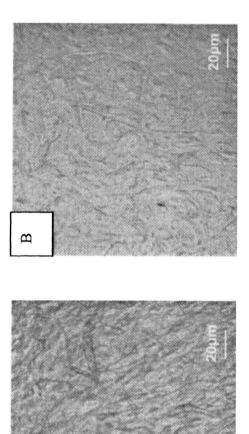

Figure 14. Optical microscopy (125 ×) showing the solid phase nanocomposite A) soybean stock as reinforcement without coating, B) soybean stock coated with ethylene-acrylic oligomer.

Acrylic oligomer emulsion coated nanofiber and PE is not completely missible; it is difficult to disperse these fibers further under compression processing. Table I shows the tensile strength and stiffness of nanocomposites in solid phase melt-mixing. Incorporation of 5% of the nanofiber improved the tensile strength without having significant effect on stiffness. Both soybean nanofiber and hemp nanofiber demonstrated reinforcing effect on the composite properties. Unlike PVA based system, the strength enhancement for solid phase polyethylene composite was not very high. It is possible that the acrylic oligomer used as a dispersion matrix for the nanocomposite acts as a low molecular weight diluent for load carrying property of the nanocomposite and the resulting low stiffness is mainly attributed to the ethylene-acrylic oligomer. Further work is required to better understand the dispersion mechanism of nanofibers into the solid phase mixing of nanocomposites.

Table I. Mechanical Properties of Nanocomposites in Solid Phase Melt mixing.

	Tensile strength (MPa)	Stiffness (CPa)
Pure PE	20-30	1-1.5
Acrylic/PE (wt%: 50/50)	7.06	0. 364
SBN/acrylic/PE (wt%: 5/45/50)	11.71	0.289
HN/acrylic/PE (wt%:5/45/50)	9.74	0.38

NOTE: SBN: soybean stock nanofiber; HN: hemp nanofiber.

Conclusion

By chemo-mechanical isolation, nanofibers having a diameter range of 50-100 nm were produced. Chemical analysis of the cellulose fiber after each stage of purification showed an increase in cellulose content and decreases the lignin content. The crystallinity of the samples increased after each stage of chemical treatment and mechanical treatment. Improved dispersion of nanofibers was achieved by adding ethylene-acrylic oligomer emulsion as a dispersant. It was observed that the tensile strength at yield of 10% nanofiber/PVA films increased by five-fold than the untreated fiber/PVA films. In solid phase nanocomposites with coated nanofibers were manufactured. Incorporation of 5% of the nanofiber improved the tensile strength without having significant effect on stiffness. The nanofibers can provide a remarkable reinforcing potential if they are uniformly dispersed in the polymer matrix. Further work is required to better understand the dispersion mechanism of nanofibers into the solid phase mixing of nanocomposites.

208

Acknowledgements

We gratefully acknowledge financial support of this study given by NSERC (Natural Sciences and Engineering Research Council of Canada).

Reference

1. Bhatnagar, A. Ms.C.F. thesis, Faculty of Forestry, University of Toronto, Toronto, ON, 2004.
2. Clowes, F.A.L.; Juniper, B.E. Plant Cells **1968**, vol. 8, 208-209.
3. McCann, M.C.; Wells, B.; Roberts K. J.Cell.Sci.**1990**, vol. 96, (2), 323-334.
4. Reiter, W.D. Trends plant Sci.**1998**, vol. 3 (*1*), 27–32.
5. Hepworth, D.G.; Bruce. D.M. Comp. Part A: Appl. Sci. Manuf. **2000**, vol. 31, 283-285.
6. Falvo, M. R.; Clary, G.; Helser, A.; Paulson; S.; Taylor; R.M. Microsc. Microanal. **1998**, vol. 4 (*5*), 504-512.
7. Hernández, E.; Goze, C.; Bernier; P.; Rubio, A. Appl. Phys. A, Materials Science and Processing. **1999**, vol. 68, 287-292.
8. Dufresné, A.; Cavaillé, J.Y.; Vignon, M.R. J. Appl. Poly. Sci. **1997**, vol. 64(*6*), 1185-1194.
9. Dufresné A.; Vignon, M.R. Macromolecules **1998**, *31*, 2693-2696.
10. Dufresné, A.; Dupeyre, D.; Vignon M.R.. J. Appl. Polym. Sci. **2000**, vol. 76, 2080-2092.
11. Stamm, A.J. Wood and Cellulose Science. The Ronald Press Company: NewYork, NY, 1964.
12. Sain, M.; Bhatnagar, A. U.S. patent 60,512,912, 2004.
13. Sain, M.; Bhatnagar, A. C.A. patent 2,437,616, 2003.

Chapter 14

Polysulfone–Cellulose Nanocomposites

Sweda Noorani[1], John Simonsen[2], and Sundar Atre[3]

Departments of [1]Chemical Engineering and Wood Science and Engineering, [2]Wood Science and Engineering, and [3]Industrial and Manufacturing Engineering, Oregon State University, Corvallis, OR 97331

Microchannel devices are a rapidly evolving process technology with a wide variety of applications. One of these is separation processes such as kidney dialysis, which promises to significantly improve treatment and patient lifestyle. Current membranes for dialysis are hollow fiber membranes typically using polysulfone (PSf). However, this geometry and membrane composition is not optimal for microchannel devices. The incorporation of cellulose nanocrystals (CNXLs) into PSf holds the potential to improve PSf membrane performance in separation devices by improving mechanical properties without loss of separation specificity or flux. However, the incorporation of CNXLs into water insoluble films without aggregation has been problematic. A solvent exchange process was developed that successfully transferred an aqueous CNXL dispersion into the organic solvent N-methylpyrrolidone (NMP), which is a solvent for polysulfone (PSf). Films were prepared from the solution of PSf in NMP with dispersed CNXLs by a phase inversion process into hot water. Films were then examined by scanning electron microscopy (SEM), optical microscopy and atomic force microscopy (AFM). The interaction between the polymer matrix and the CNXL filler was studied by means of thermogravimetric analysis (TGA), which suggested a close interaction between the polymer and filler at the 2% filler loading.

209

Introduction

Polymeric composite materials filled with nanosized rigid particles have attracted great interest in recent years (1, 2). In addition, the use of nanoparticles of natural origin has also been of interest (3). The advantages of natural fillers are their low density, renewable character, high specific strength and biodegradability associated with the highly specific properties of nanoparticles (4). Cellulose, a semi-crystalline polymer, is one of the most abundantly available polymers on earth. Cellulose is present in virtually all plants and is also produced by certain varieties of bacteria and some sea animals (5). Depending on its origin, the molecular weight and the degree of polymerization vary. The crystalline portions of cellulose can be obtained by acid hydrolysis (6, 7) and are called cellulose whiskers or cellulose nanocrystals (CNXLs). CNXLs prepared in our lab were obtained from cotton and had a length of 200-350 nm, cross section of 5 - 20 nm and aspect ratio between 10-70 nm. The density of CNXL calculated from X-ray diffraction data is 1.566 g/cc (6). The stiffness of CNXLs has been measured at 145 GPa (8). The strength has not been measured directly, but theoretical estimates are about 7500 MPa (9).

Polymer nanocomposite films incorporating CNXLs are being developed in our research group specifically to be utilized in microchannel-based devices for separation technologies. Such devices require thin walled membranes with high stiffness and flux. Rapid separation times and scalable throughputs are made possible by several microchannel features such as high mass transfer rates, short residence time distribution, control of the diffusion length, and mixing (10, 11). These benefits translate into a small device size which is ideal for integration into portable and distributed systems (12). Figure 1 shows some prototype polysulfone (PSf) microchannel devices designed for bioseparations (13). One potential application for such microchannel devices is kidney dialysis. By reducing the size and cost of the device, this technology offers the potential to allow dialysis patients to perform the operation at home, resulting in significantly improved patient treatment and lifestyle (14).

Incorporation of CNXLs as a reinforcing material in a PSf matrix has the potential to provide the necessary mechanical and separation properties for an optimal microchannel separation device. As a preliminary study, CNXLs were incorporated into a PSf matrix and the morphology and thermal properties of these films were investigated.

Experimental

Polymer Matrix

The PSf resin (UDEL P-1700) was donated by Solvay Advanced Polymers (Alpharetta, GA). The melt flow index is listed as 6.5 g/10 min, tensile strength

Figure 1. Prototype microchannel device for bio-separations

as 70.3 MPa, tensile modulus 2.8 GPa, elongation at break, 50-100%, specific gravity 1.24. The polymer was received in the form of pellets and was used without further purification.

CNXL Production

CNXLs were prepared by acid hydrolysis of pure cotton cellulose (Whatman #*1* filter paper), ground to pass 20 mesh, at 45 °C for 45 minutes in 65% sulfuric acid. The solution obtained was centrifuged and neutralized with 5% $NaHCO_3$ until the solution pH was 6 - 9. This solution was then sonicated using a Branson Sonifier 250 (Branson Ultrasonics Corp., Danbury, CT) for about 10 minutes. After sonicating, the solution was filtered in an Ultrasette™ Tangential flow ultrafilter, 300 kD MWCO, from Pall Corp. (Ann Arbor, MI). Filtration was continued until the final conductivity of the ultrafilter permeate was < 5 μS/cm.

Nanocomposite Processing

A solvent exchange process was used to transport the CNXLs from water to the organic solvent N-methylpyrrolidone (NMP). CNXL aqueous dispersion and NMP were mixed together. The amount of NMP added depended on the percentage CNXL required in the final composite. The water was removed via a Büchi Rotovapor R110 (Flawil, Switzerland) rotary evaporator. The temperature of water bath was adjusted to 50-80 °C, a vacuum was applied, and the process continued until all the water was removed from the mixture as determined by the weight of the remaining dispersion. The dispersion was then sonicated for about 10 minutes. PSf was added to this solution mixture and stirred until the polymer was completely dissolved. The dispersion was sonicated again. This dispersion was used to make films by the phase inversion process: Solutions were cast on a glass plate and immersed into hot water (70 °C). The films were left in the hot water bath for at least 24 h to remove the remaining solvent from the films. The percentage of CNXLs in the films was varied between 0 and 16 wt %. The thickness of the final films was ~ 20 μm and was measured by calibrated optical microscopy.

Thermogravimetric Analysis (TGA)

Thermogravimetric measurements were carried out with a TA Instruments Q500 thermogravimetric analyzer (New Castle, DE). The temperature range was room temperature to 600 ° C at a 10 ° C/min ramp under nitrogen flow.

Atomic Force Microscopy (AFM)

A Digital Images Dimension 3100 atomic force microscope (Veeco Instruments Corp., Santa Barbara, CA) was used to obtain height and tapping mode images of the surfaces of the films.

Scanning Electron Microscopy

An AMRay 1000 scanning electron microscope (SEM) using 10kV secondary electrons was used to image the morphology of the nanocomposite films. The specimens were frozen in liquid nitrogen, fractured, mounted, coated with gold/palladium and the fracture surface was observed at a 300 angle.

Optical Microscopy

Images were obtained on an Eclipse E400 (Nikon USA, Melville, NY).

Results and Discussion

Morphology of Films

The fracture surface of the unfilled PSf appeared to be a smooth leaf-like structure (Figure 2A). The fracture surface of films containing CNXLs showed a radically different topology. At the 2% loading, the fracture surface of the film was rough and non-uniform with ~ 1 μ fracture zones (Figure 2B). Clearly the presence of the nanoparticles was altering the fracture behavior of the material. Increasing the CNXL loading to 11% gave a fracture surface similar to the 2% sample, except the fracture zones were somewhat smaller and more numerous (Figure 2C). This would be expected if the nanoparticles were altering the crack propagation process since more nanoparticles should give greater crack trajectory modifications. At the 16% CNXL loading level, the fracture surface was extremely rough, uneven, and showed apparent domains of agglomerated CNXLs (Figure 2D). This loading of CNXLs evidently exceeds the ability of the PSf to wet the filler surface, giving rise to the generally "cheesy" behavior.

Figures 3 (A - C) show the polarized optical micrographs of PSf films with 0%, 2% and 11% filler loadings respectively. At lower filler loadings (2%) the

214

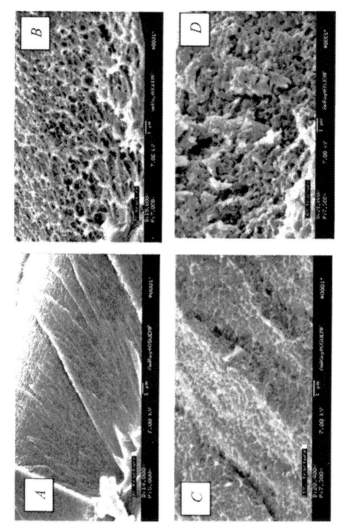

Figure 2. Fracture surfaces of PSf film filled with A) 0%; B) 2%) 11%; D) 16% CNXLs.

215

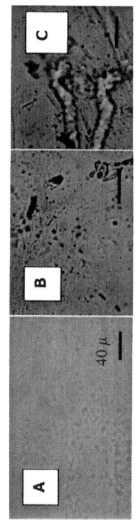

Figure 3. Polarized optical micrograph of PSf film containing A) 0%; B) 2%; and C) 11% CNXLs.

CNXL is less agglomerated while at higher filler loadings (11%) the CNXL is highly agglomerated. Note that the individual CNXLs are not visible in an optical microscope, being smaller than the wavelengths of visible light. However, the small particles still diffract and scatter light and produce out-of-focus "spots" in optical images. Agglomerates larger than ~ 500 nm are easily visible and are apparent in Figures 5 B and C, especially C, where large agglomerates are apparent.

AFM images also show agglomeration at the higher CNXL loadings (Figure 4). These images were acquired using tapping mode AFM. The intent was to asses the feasibility of "seeing" or "sensing" the CNXLs imbedded in the film using tapping mode AFM and phase imaging. The phase image responds to a number of variables and currently there is no theory that allows the analysis or prediction of phase images from first principles (15). Nevertheless, phase imaging has become a useful tool for the atomic force microscopist due to its versatility and image quality. One of the variables which effects the phase image is the hardness of the material being imaged. Typically the harder a substance is the brighter it displays in a phase image. Thus regions of "hard" and "soft" materials can be discerned in phase images. Since the CNXLs are much harder than the PSf matrix, they should be visible in the phase image, even though they are not on the surface. It was expected that as the force, or "hardness" of the tapping increased, more deeply imbedded CNXLs should appear in the phase image. There is some evidence to support this hypothesis. Comparing Figure 5 with Figure 4C, one can find additional bright areas which may be more deeply imbedded CNXLs. The only difference between the images is the level of the setpoint, which determines the degree of "hardness" of the tapping mode. Thus we conclude that this technique, while certainly not quantitative, does hold some merit for qualitatively analyzing polymer nanocomposite films.

Thermogravimetric analysis

Unfilled PSf films and composites filled with various CNXL loadings were tested using TGA. The degradation of pure PSf was observed to begin at ~500 °C and that of pure CNXLs at 250 °C (Figure 6). The 11% CNXL loading showed a bimodal distribution with the expected ~250 °C CNXL degradation step and the ~ 500 °C PSf step occurring relatively independently at the degradation temperatures expected for the pure components. However, the 2% CNXL loading showed a broad CNXL degradation step shifted to a higher temperature (Figure 6). This may indicate that the CNXL and PSf components are associated, altering the thermal degradation of the CNXL. Thus at CNXL loadings > 2% there may be agglomeration and the agglomerated CNXL phases may be degrading independently of the PSf matrix. This leads to the supposition that the surface area of the CNXL component is large enough that at some CNXL loading between 2% and 11% the PSf fails to fully "wet" the surface.

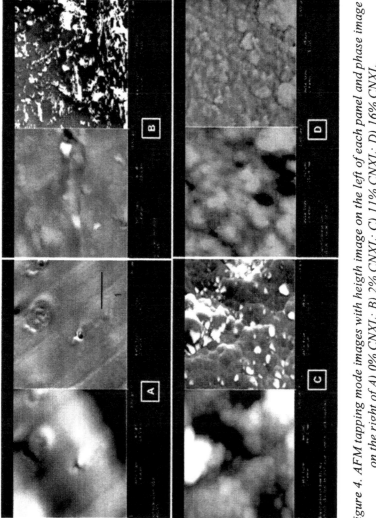

Figure 4. AFM tapping mode images with heigth image on the left of each panel and phase image on the right of A) 0% CNXL; B) 2% CNXL; C) 11% CNXL; D) 16% CNXL.

217

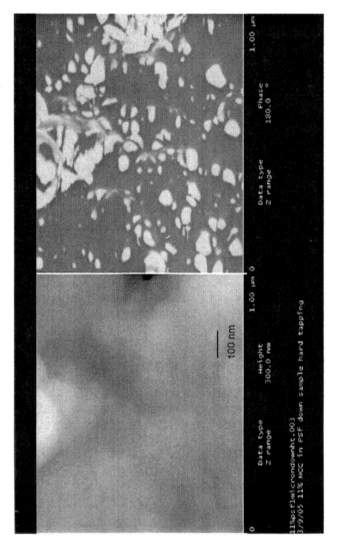

Figure 5. AFM image of 11% CNXL -filled PSf using the hard tapping technique. Compare to Figure 6C.

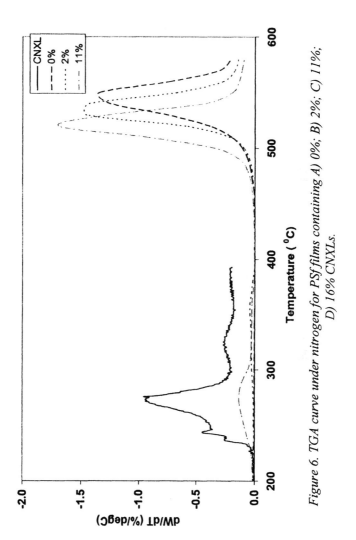

Figure 6. TGA curve under nitrogen for PSf films containing A) 0%; B) 2%; C) 11%; D) 16% CNXLs.

Conclusions

CNXLs were incorporated into a PSf matrix by a novel solvent exchange process. A solid composite film was obtained via phase immersion casting. Microscopic observations revealed that the CNXLs were dispersed well at lower filler loadings. SEM indicated that the incorporation of CNXLs radically altered the fracture behavior of the composites. In addition, the 16% CNXL loading appeared to be highly agglomerated and incompletely wet by the matrix. AFM images indicated that "hard" tapping with phase imaging was a promising technique for visualizing films such as these. TGA suggested strong PSf/CNXL interactions at 2% filler loading and agglomeration or poor PSf/CNXL interactions at 11% and higher filler loadings.

References

1. Favier, V.; Canova, G.; Shrivastava, S.; Cavaille, J. *Polym. Eng. Sci.* **1997**, *37*, (*10*), 1732-1739.
2. Morin, A.; Dufresne, A. *Macromol.* **2002**, *35*, 2190-2199.
3. Ruiz, M.; Cavaille, J.; Dufresne, A.; Gerard, J.; Graillat, C. *Comp. Interf* **2000**, *7*, (*2*), 117-131.
4. Samir, A.; Said, M. A.; Alloin, F.; Sanchez, J.-Y.; Dufresne, A. *Polymer* **2004**, *45*, (*12*), 4149-4157.
5. Ward, K.; Ott., E.; H.M. Spurlin; M.W. Grafflin, *Occurence of Cellulose.* Interscience Publishers: NY, 1954; Vol. Part I, p 9-27.
6. Battista, O. A., *Microcrystal Polymer Science.* **1975**.
7. Revol, J.-F.; Giasson, J.; Guo, J.-X.; Hanley, S. J.; Harkness, B.; Marchessault, R. H.; Gray, D. G., *Cellul.: Chem., Biochem. Mater. Aspects* **1993**, 115-22.
8. Eichhorn, S. J.; Young, R. J.; Davies, G. R., *Biomacromol.* **2005**, *6*, 507-513.
9. Marks, R. E., *Cell wall mechanics of tracheids.* **1967**.
10. Martin, P. M.; Matson, D. W.; Bennett, W. D., *Chem. Eng. Comm.* **1999**, *173*, 245-254.
11. Martin, P. M.; Matson, D. W.; Bennett, W. D., *J. Vac. Sci. Tech. A.* **1999**, 17, 2264-2269.
12. Wegeng, R. S.; Drost, M. K.; Brenchley, D. L. In *Process Intensification Through Miniaturization or Micro Thermal and Chemical Systems in the 21st Century,* 3rd International Conference on Microreaction Technology, Frankfurt, Germany, 1999; Frankfurt, Germany, **1999**.
13. Urval, R. Powder Injection Molding of Multi-scale Devices. M.S., Oregon State University, Corvallis, OR, **2004**.
14. Curtis, J. *Dialys. Transpl.* **2004**, *33*, (*2*), 64-71.
15. Cohen, S. H.; Lightbody, M. L., *Atomic Force Microscopy/Scanning Tunneling Microscopy 2.* Plenum Press: New York, **1997**.

Chapter 15

Bacterial Cellulose and Its Nanocomposites for Biomedical Applications

W. K. Wan[1,2], J. L. Hutter[1,3], L. Millon[1], and G. Guhados[2]

[1]Graduate Program in Biomedical Engineering, Departments of [2]Chemical and Biochemical Engineering, and [3]Physics and Astronomy, The University of Western Ontario, London, Ontario, Canada

Bacterial cellulose is a natural nanomaterial produced by strains of the bacterium *Acetobacter xylinum* in static, shaken and agitated cultures. With fiber dimension of ~30 nm and being very hydrophilic, it is ideally suited for a broad range of biomedical applications. The bacterial cellulose fiber has a high degree of crystallinity. Using atomic force microscopy for direct nanomechanical measurement, we have determined that these fibers are very strong and when used in combination with other biocompatible materials, produce nanocomposites suitable for use in medical device applications. In particular, nanocomposites consisting of bacterial cellulose and poly(vinyl alcohol) have been investigated. A broad range of mechanical properties control was demonstrated. Materials with properties closely mimic that of cardiovascular tissues were prepared. These materials are potential candidates as synthetic replacements of natural cardiovascular tissues.

Introduction

Cellulose is one of the most abundant naturally occurring polymers on Earth. It has many properties, including its polyfunctionality, multichirality, hydrophilicity, and biocompatibility (*1*), that make it valuable as both an industrial and biomedical material. Bacterial cellulose has the added advantage of not containing impurities that need to be removed by harsh chemical treatment (*2*). Moreover, it is a natural nanomaterial typically occurring as fibers of diameter ~3 nm grouped into bundles of diameter ~50 nm, thus resulting in a high surface-to-volume ratio (*3*).

Bacterial cellulose is produced by strains of the bacterium *Acetobacter xylinum*. In nature this bacterium is typically found on decaying fruits and vegetables (*4*). It is also common in fruit juices, and can be a source of problems in vinegar and the production of alcoholic beverages. The exact reason why *A. xylinum* expends a significant portion of its nutrients and energy to produce bacterial cellulose is not known, but it is believed that the bacterial cellulose serves to hold the bacteria in an aerobic environment (*5, 6*), and protect it from ultra violet light and predators (*7*).

Since A. J. Brown (*8*) discovered and identified bacterial cellulose production in static culture, there has been much interest in its commercial exploitation. The most common use of bacterial cellulose to date is as an indigenous food in the Philippines called nata-de coco, which is made by fermenting 1-cm thick sheets of bacterial cellulose with coconut water. This gel sheet is then cut into cubes, immersed in syrup, packaged and sold. Bacterial cellulose is also used as a wound dressing material (*2*) and for acoustic diaphragms in high-quality headphones (*9*). However, there are many potential applications for the use of microbial cellulose that spur research at a very rapid pace. These include: bacterial-cellulose-containing conductive carbon film (*10*); high quality paper for bank notes and bible pages (*9*); separation membranes (*11*); dispersing aids (*3*); thickeners (*3*); dietary fiber food (*3*); composites for strength improvement of polymers (*12*); fine crystalline product when ground with paint, pharmaceuticals and cosmetics (*12*); coatings for hydrophobic polymers to increase their hydrophilicity (*12*). Klemm *et al.* have developed vascular conduits based on bacterial cellulose of inner diameter of 3 mm that have been successfully implanted in rats (*13*). Bacterial cellulose has also been reported to have excellent properties and mechanical strength for potential scaffold material for tissue engineering of cartilage (*14*).

Structure of Bacterial Cellulose

Brown (*8*) first discovered cellulose production by bacteria. He noticed that certain strains of *Acetobacter* produced a white gelatinous pellicle on the surface of a liquid medium and determined that the structural material of this pellicle was chemically identical to cellulose. Other strains of *Pseudomonas,*

Achromobacter, Alcaligene, Aerobacter and *Azotobacter* were also found to produce cellulose (*12*) when grown in the appropriate medium.

Early electron microscopy studies of bacterial cellulose pellicles showed that they consisted of a thick network of ribbons 10 nm thick and 500 nm wide (*15*). These ribbons were themselves comprised of smaller fibers 20 – 25 nm in diameter, similar to microfibrils found in plant cell walls. However, more than a decade later, Ohad *et al.* (*16*) showed that metal shadowing used to enhance the contrast of the electron micrograph caused metal deposits on the fibers that resulted in a gross overestimation of their size. They extrapolated from shadowing experiments that the actual width of a bacterial cellulose microfibrils was only 3 nm. Colvin (*17*) showed that negative staining, one of the methods used to refute the metal shadowing experiments, could underestimate the fiber width. However, since then, further work has supported Ohad *et al.* (*16*) and confirmed that the width of a cellulose microfibril is indeed between 2 and 3 nm (*18*) and the width of a bundle is less than 130 nm. Crystallographically, bacterial cellulose belongs to Cellulose I, the same as natural cellulose of plant origin (*9*).

Various researchers have measured the degree of polymerization of bacterial cellulose (*19 - 23*). Many have noticed a decrease in the average degree of polymerization as fermentation progresses and have attributed this decrease to the production of cellulase by the bacteria. However, Marx-Figini (*19*) theoretically predicted the reduced average degree of polymerization based on chain propagation kinetics.

Production of Bacterial Cellulose

Static Culture of Acetobacter Xylinum

Growing *A. xylinum* in a static liquid medium results in the formation of a tough cellulose pellicle at the surface of the liquid, in which the majority of the cells are immobilized (*24*). Oxygen is essential for both cellulose production and cell division. Schramm *et al.* (*25*) showed that cellulose was not produced when *A. xylinum* was grown under nitrogen and supplied with alternative electron acceptors. An abundant supply of oxygen at the surface of the medium leads to a rapid synthesis of cellulose. The formation of the pellicle, which is essentially a mesh of cellulose fibers, traps *A. xylinum* cells at the surface of the liquid medium. Another major effect of the pellicle is that it provides a significant barrier to the diffusion of oxygen into the liquid medium. In fact, only cells close to the surface of the pellicle produce cellulose at an appreciable rate (*26*). Additionally, Fontana *et al.* (*2*) found that the cellulose pellicle is actually formed from a large number of thin parallel cellulose sheets, with the more recently formed layers nearer the film-air interface (*26, 27*).

The effect of various parameters on static culture of *A. xylinum* has been studied by a number of researchers to determine if the rate of production of bacterial cellulose can be improved. Many of these studies produced conflicting results because of inconsistencies in analytical techniques, strain of bacteria used and varying culture time (*28 - 36*)

An alternative form of static culture that has been developed recently uses rotating disks partly submerged in a liquid medium suitable for bacterial cellulose production with high production rate (*37*).

Shaken and Agitated Culture of A. *xylinum* Selection of Strains

Many static culture strains of *A. xylinum* seem to produce less cellulose in shaken culture (*31*). Some have speculated that the reduction in cellulose production is a result of the increased proliferation of cellulose-negative mutants of *A. xylinum* in shaken culture. Insertional sequences (ISs), which can cause genetic instability were discovered in cellulose-producing *Acetobacter* strains (*14*). Strain selection is therefore important to the successful production of bacterial cellulose in shaken and agitated cultures (*35, 36, 38*)

Factors Affecting Cellulose Production in Shaken and Agitated Culture

Dudman (*28*) was the first to report the production of cellulose in submerged agitated cultures. He compared two strains: *Acetobacter acetgenum* EA-I and *A. xylinum* HCC B-155. *A. xylinum* BPR 3001A, which is similar to BPR 2001, produces bacterial cellulose particles of many sizes (10 μm to 1 mm) and shapes (spherical, ellipsoidal, satellate or fibrous) in agitated culture.

Efforts to improve the productivity of *A. xylinum* in agitated cultures have been reported (*39 - 45*). The principal approach is by the addition of viscosity modifiers to modulate the shear effect due to shaking and agitation. *A. xylinum* also produces a water-soluble polymer namely, acetan at different shaking speeds, thereby reducing the product cellulose yield (*46*).

Recently, an isolate, *Acetobacter sp. A9*, was found to produce cellulose in substantial amounts under both static and shaking conditions. Productivity is further increased with the addition of ethanol (*47*).

Other factors affecting bacterial cellulose production in both shaken and agitated cultures include carbon source (*24, 32, 48 - 50*), nitrogen source (*50-52*), vitamins and minerals (*49*) and dissolved oxygen concentration (*28, 29, 31, 53*).

Current Study

Bacterial cellulose is a hydrophilic, natural nanomaterial that is ideally suited for a broad range of biomedical applications. In particular, it can be used in the preparation of hydrophilic, biocompatible nanocomposites with controlled mechanical properties for replacement medical devices (*13*). We report results of our studies on structural and mechanical properties of these fibers and their polyvinyl alcohol (PVA) nanocomposites.

Materials and Methods

Bacterial Cellulose Preparation

Bacterial cellulose was produced by *Acetobacter xylinum* BPR 2001 (ATCC #700178) in shaken and agitated cultures. Inoculum was grown at 175 rpm and 28°C for 3 days, after which shake flasks containing 100 mL of sterile medium as per Joseph *et al.* (*45*) were inoculated and maintained at 28°C to produce cellulose. The product was purified by processing with 1% NaOH solution for 30 minutes in a boiling water bath to lyse the bacteria. The presence of polyacrylamide–co–acrylic acid (PA) in shake flasks resulted in increased cellulose yields.

Samples for atomic force microscopy were prepared by spin coating 50 μL of 0.36 wt% bacterial cellulose suspension onto an atomic force microscope (AFM) calibration grating (NanoAndMore, Germany) consisting of ridges separated by trenches 1.5 μm across and 1.0 μm deep. This typically resulted in clumps of bacterial cellulose with some isolated fibers at the periphery.

Bacterial Cellulose Characterization

Fourier transform infrared spectroscopy (FT-IR) spectra were recorded using a Bruker Vector 22 spectrometer. Samples were analyzed in absorbance mode, with accumulation of 32 or 64 scans.

Samples of bacterial cellulose were analyzed using a Rigaku DMX-2 X-ray powder x-ray diffractometry, with Cu-K alpha radiation. Data were collected from 5° to 40° at increments of 0.05° with 2 s time intervals. The degree of crystallinity was calculated according to Eichhorn *et al.* (*54*) and Horii *et al.* (*55*).

Atomic Force Microscopy was performed using a Multimode AFM with Nanoscope IIIa controller (Veeco Instruments). Bacterial cellulose samples were imaged in air using contact mode with Si_3N_4 cantilevers of nominal spring constant 0.5 N/m (MLCT Microlever F, Veeco Instruments). The AFM images allowed isolated fibers spanning the trenches to be located. Once such fibers were identified, the AFM was used to measure their mechanical properties.

Nanomechanical Measurements

Details of our measurements of the mechanical properties of bacterial cellulose have been described previously (*56*). Briefly, mechanical testing was performed using the AFM in force-volume mode, in which force spectra (i.e., cantilever deflection as a function of vertical displacement of the piezoelectric scanner) are recorded at an array of positions in the vicinity of a suspended fiber. Slow vertical ramp speeds of 1 μm/s (500 nm ramps acquired bi-directionally at rates of 1 Hz) were used to avoid viscoelastic effects. To avoid plastic deformation, we limited the applied force to a maximum of 90 nN via a trigger set-point.

Forces applied during force spectra are determined from the cantilever displacement and its measured spring constant (1.03 ± 0.05 N/m in this case, as measured by the thermal noise method (*57*) and verified by the added mass method (*58*)). Since both the sample displacement and cantilever deflection are recorded, the sample deformation can be calculated as the difference in these quantities after contact is achieved. Thus, the sample deformation can be measured as a function of applied load.

Data Analysis and Young's Modulus Determination

To determine the mechanical properties of a suspended fiber from measurements of its deflection in response to an applied load, a model for the expected deflection is needed. The deflection δ of a clamped, suspended beam of constant cross-section and length L subjected to pure bending by a concentrated load F applied at a point a relative to one of its ends is, at the point where the force is applied, (*59*)

$$\delta(a) = \frac{F}{3EI} \left[\frac{a(L-a)}{L} \right]^3 \qquad (1)$$

where I is its area moment of inertia.

In order to apply this model to measurements of suspended fibers, a number of assumptions must be justified. For instance, this model assumes that the beam is clamped securely to the substrate, so that the slopes at the ends of the suspended portion are zero. This cannot be directly verified because the AFM measures the deflection of only one point at a time the deflection at nearby points, and hence the slope at the endpoints, is not accessible. Indirect evidence of secure anchoring includes the observation that repeated scanning in a direction perpendicular to the beam axis does not result in lateral displacement of the supported portions of the fibers, even when displacement of the *suspended* portion is obvious.

Equation 1 also neglects other potential deflection mechanisms, such as that due to internal shear. For an isotropic material, shear deflection at the center of a fiber of elliptical cross-section and vertical thickness h is only as important as the bending deflection when $h \approx L/f_s \sqrt{6}$, a condition never approached for samples in this study (here, f_s is a shape factor of order unity) (59). However, there is no *a priori* reason to assume that these bacterial cellulose fiber bundles are isotropic. One might for instance expect shear, in the form of slip between adjacent bundles, to play a role. If such shear can be characterized by an effective shear modulus G, we expect an additional deflection of approximately (59)

$$\delta_{\text{shear}}(a) = f_s \frac{F}{GA} \frac{a(L-a)}{L} \qquad (2)$$

where A is the cross-sectional area of the fiber. As our data does not support such a correction for small values of G, we conservatively assume $G=E/2(1+v)$, as appropriate for an isotropic material.

Also neglected is the possibility of deflection due to stretching of the fiber – i.e., "cable" behavior. However, a clear signature for this would be a nonlinear dependence of cantilever deflection on sample displacement, which is not observed. Moreover, a simple estimate of the maximum strain due to stretching, based on the measured deflection at the fiber center

$$\varepsilon_{\text{bend}} = \frac{2\delta^2}{L^2} \qquad (3)$$

predicts that this stretching strain will only become significant relative to the maximum bending strain at the fiber center (determined from the maximum fiber curvature)

$$\varepsilon_{\text{bend}} = \frac{12h\delta}{L^2} \qquad (4)$$

for deflections approaching six times the fiber thickness. As typical deflections in this experiment are approximately 50 nm, whereas fiber thicknesses ranged from 30 – 60 nm, we are able to neglect this effect.

Bacterial Cellulose (BC) – Polyvinyl Alcohol (PVA) Nanocomposite

Preparation

The PVA used (Sigma-Aldrich Canada Ltd.) had an average molecular weight (M_w) in the range of 124,000-186,000, and was 99+% hydrolyzed. Aqueous solutions of 10 wt% PVA were prepared according to an established procedure (*60*). Suspensions of BC in distilled water were produced in our lab (*45, 56*). Depending on the required composition, dry PVA was added to the BC suspension with or without extra distilled water. This mixture was heated following the same procedure as for the PVA solution (*60*). The PVA and PVA–BC nanocomposites were prepared from their solutions by a repeated freezing and thawing process (0.1°C/min) between 20 and -20°C, with a 1 hour hold at the low-temperature extreme (*60*). Samples were cut into 10x25 mm specimens (n=10) for tensile testing.

Porcine Aorta and Aortic Heart Valves Samples

A total of 6 fresh porcine hearts with intact aortas and aortic heart valves were obtained from an abattoir. The aortas were cut axially and flattened into a rectangular shape. These samples were cut along both circumferential and axial directions for tensile testing (10x25 mm for each direction, n=8).

For the aortic heart valve cusps, a total of 18 cusps were dissected, 9 samples each for the circumferential and radial directions. Due to sample size limitations, they were cut into either 5x15 mm (for circumferential) or 10x10 mm (for radial) samples. All tissue samples were stored and tested in Hank's saline solution.

Tensile Test

The testing equipment included a uniaxial hydraulically powered material testing system (MTS Bionix 858) with a 1 Kg load cell, as described previously (*60*). A custom designed Mitutoyo gauge thickness tester was used to measure the sample thickness. Testing was carried out in either distilled water (PVA and PVA-BC samples) or Hank's saline solution (tissue samples) at 37°C. All specimens were secured with custom designed tissue grips (~10 mm grip-to-grip distance).

Data Analysis

The data obtained was in the form of load-extension, which was then converted into engineering stress-strain using the sample thickness and the initial gauge length after preconditioning.

The stress-strain data for both PVA and PVA-BC samples, as well as for tissue, is non-linear. For the PVA samples, the stress-strain data was fitted by (*61, 62*):

$$\sigma = y_0 + A \exp(B\varepsilon) \qquad (5)$$

where σ is stress, ε is strain, and y_o, A, and B are free parameters. The elastic modulus is calculated as the first derivative of eq 5.

For PVA-BC, aorta and aortic heart valves, a better fit was obtained from (*62, 63*):

$$\sigma = y_0 + A \exp(B\varepsilon) + C \exp(D\varepsilon) \qquad (6)$$

where y_o, A, B, C, and D are curve fitting parameters. Again, the elastic modulus was calculated as the first derivative.

Results and Discussion

Bacterial Cellulose Production

Bacterial cellulose was produced in baffled shake flasks according to Joseph *et al.* by *Acetobacter xylinum* BPR 2001 (*45*). Previous reports showed that *Acetobacter xylinum* BPR 2001 produced high yields of cellulose relative to other strains using fructose-corn steep liquor (CSL) based medium. A fructose concentration of 2% was used to shorten the fermentation time as well as to reduce the deleterious effects of high bacterial cellulose concentration due to mixing and mass transfer. Our results show that the addition of polyacrylamide (3%) results in a significantly higher yield: 4.73 ± 0.32 g/L (n=3) compared to 1.69 ± 0.41 g/L (n=3) without its addition.

The effect of thickener in fermentation has been studied previously by Ben-Bassat, A. *et al.* (*43*), Joseph *et al.* (*45*) and Moo-Young, M. *et al.* (*44*). The addition of thickeners results in interesting rheological changes in the medium.

The main advantage of polyacrylamide is its availability over a wide range of molecular weights, that can be used according to our requirements. Studies of shake flask culture by Joseph *et al.* (*45*) reported that the addition of 3% polyacrylamide resulted in controlled product morphology and high cellulose yield. Our results are in agreement with Joseph *et al.*

Mechanical Properties of Bacterial Cellulose Nanofibers

Figure 1 shows a force-volume image of a bacterial cellulose fiber bundle spanning a 1.5 μm gap. This bundle was measured (via images of higher resolution) to have a semi-minor axis of 17 nm and semi-major axis of 55 nm (appropriately corrected for tip shape). The apparent width of the fiber is greater in the gap than on the rigid surface for the same reason that the edges of the gap appear sloped: convolution with the pyramidal AFM tip. To avoid erroneous force estimates due to contact between the suspended fiber and the sides of the tip, we chose only points (indicated by dots in Figure 1) along the midline of the fiber (indicated by a line) for further analysis.

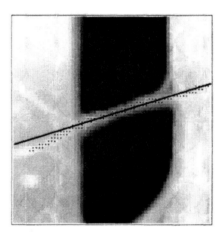

Figure 1. Force-volume height image of a bacterial cellulose fiber suspended across a 1.5 μm gap. The dots show positions where force spectra were collected for use in mechanical modeling.

The slopes of force spectra measured at points selected from Figure 1 are plotted as a function of position along the fiber in Figure 2. For positions at which the fiber is supported by the underlying substrate, the slope reaches its maximum deflection of 1, indicating that no sample deformation occurs. For points between the supports, the ability of the fiber to be depressed results in smaller slopes. The solid curve indicates a fit to the model described previously and yields a Young's modulus of 77 GPa.

Some deviation from the model fit in Figure 2 is evident, most noticeably near the right-hand edge of the supported fiber. Examination of this region in the height image reveals that the fiber branches into two smaller bundles, only one of which follows the line of the main fiber. Naturally, the branch that is measured can be expected to be somewhat weaker than the complete bundle because forces applied to it do not cause the entire bundle to bend.

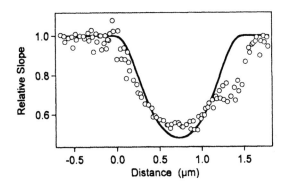

Figure 2. Slopes of force spectra measured at positions along a suspended fiber as a function of position along the fiber. The solid curve indicates the best fit to the model function used to determine the Young's modulus.

The cubic dependence of the measured Young's modulus on the vertical height of the fiber and the linear dependence on the less easily measured width result in a large uncertainty (and hence scatter) in its measured value. Moduli ranging from 45 – 140 GPa were determined from measurements of eight fibers. Since no dependence on fiber-bundle dimensions (minor axis, major axis, or the geometric mean of these quantities) could be discerned, we assume that the properties of all eight fibers are similar, and take the mean value of the Young's moduli determined to characterize bacterial cellulose. We find a value of E 84 ± 16 GPa for mean radii ranging from 31 – 72 nm, where the uncertainty is the sum of the standard error of the mean (12 GPa) and an additional 4 GPa due to the uncertainty in the spring constant calibration (which does not contribute to the scatter in the results).

Direct measurements of the mechanical properties of nanomaterials are at an early stage. Previous AFM determinations of the Young's modulus of nanofibers were based on measurements of deflection at a single point near the center of the fiber, which does not provide validation for the elastic model employed (*64*). In this study, we measured the deflection at a series of points along the fiber axis, allowing its dependence on position to be compared with the predictions of an elastic model. The fits are satisfactory, though a slight discrepancy, most likely due nonuniformity in the fiber cross-section (i.e., branching) can be seen in the case shown, demonstrating the importance of model validation. In this study, justification of the clamped-beam model is provided by observations that the portion of the fiber on the substrate remains stationary, even when lateral forces are applied, and by the quality of the model fit to the data.

Previous studies of the mechanical strength of cellulose indicate a strong dependence on crystallinity, which has been modeled by considering the fiber to be a composite of crystalline and amorphous phases. Two models, defined by a parallel (Voigt model) or series (Reuss model) arrangement of the crystalline phase, limit the range of Young's moduli possible. With assumptions for the Young's moduli of amorphous and crystalline cellulose (5 MPa and 128 MPa, respectively), Eichhorn *et al.* developed a relationship between the Young's modulus and the crystalline fraction of cellulose samples, and demonstrated its success for microcrystalline cellulose, flax and hemp (*54*). According to this model, the value for our bacterial cellulose samples with 60% crystallinity should lie between 12 GPa (Reuss model) and 74 GPa (Voigt model). The value of 84 ± 16 GPa determined here implies that the crystalline regions in the bacterial cellulose fibers are predominantly arranged in parallel, as in the Voigt model, and as expected from electron microdiffraction experiments (*65*). This value is the first determination of the Young's modulus of cellulose at nanometer length scales and, hence, is the first direct measurement of a cellulose nanofiber and represents one of the highest values for cellulose at a similar crystallinity.

BC – PVA Nanocomposites

To facilitate the presentation of data in the Figures and Tables, PVA is denoted by (P) and bacterial cellulose as (BC). The actual concentration of each component is given on a weight percentage basis after the letter representing that component.

For a biomaterial to be used as a tissue replacement, it is important to ensure a good match of the mechanical properties of the implanted device to the surrounding tissues (*66*). PVA hydrogels have many properties that make them good candidates for use in biomedical applications (*67, 68*). However, even though PVA shows nonlinear exponential tensile properties similar to aortic root and arteries, the mechanical properties deviate from those of tissue after 30% strain (*60*). PVA-BC nanocomposite hydrogels, on the other hand, have improved mechanical properties more closely matching those of cardiovascular tissue.

For soft tissue replacement applications, it is important to note that cardiovascular tissues, such as artery or heart valve leaflet, are also composite materials, with elastin and collagen as the main load-bearing components, and also containing vascular smooth muscle surrounded by ground substance. In general, the tensile properties of cardiovascular tissues show an initial low modulus portion, where elastin stretches while the slack in the collagen fibers is reduced. Collagen only contributes to the vessel wall tension as the vessel is further stretched, and the collagen fibers become fully stretched (*61, 69, 70*). Thus, it is understandable that a composite material might be required to provide a match of the mechanical behavior of such tissues.

PVA can itself be described as a two component composite material. It is known that PVA crystallites are formed through localization of polymer-rich regions, with a high degree of organization, surrounded by water-rich amorphous regions. The main inter- and intra-molecular interactions are hydrogen bonds and van der Waals forces. PVA mechanical properties also exhibit a non-linear stress-strain relationship similar to that displayed by cardiovascular tissues (*60, 70-73*), with increasing stiffness as a function of thermal cycles (Figure 3), as previously described (*60, 71*). A statistically significant difference (P<0.05) was observed among the six cycles.

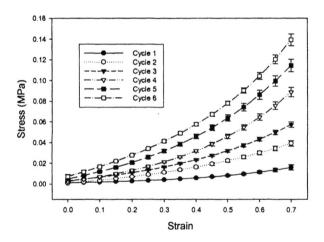

Figure 3. Stress-Strain curves for 10% PVA hydrogels (cycles 1 to 6).

In the PVA-BC nanocomposite, BC acts as third component. BC is very hydrophilic, and due to its large aspect ratio is expected to experience strong hydrogen bonding with the PVA polymeric chains. The full range of tensile properties obtained with PVA-BC nanocomposites in the BC concentration range of 0.15 – 0.6% (cycle 6) can be seen in Figure 4. The error bars are omitted for clarity. As well, the specific effect of BC at a constant PVA concentration and the effect of PVA at a constant BC concentration are shown in Figures 5 and 6 respectively. In Figure 5, when ANOVA was applied to the stress of all composites at a strain of 65%, all groups were found statistically different (P<0.05). In Figure 6, when ANOVA was applied to the stress of all

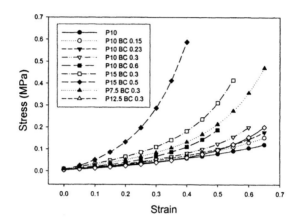

Figure 4. Stress-Strain curves for all 9 PVA-BC nanocomposites and the PVA control (cycle 6).

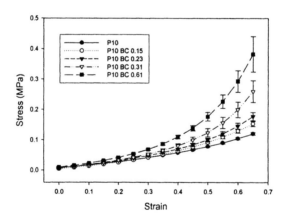

Figure 5. Stress-strain curves of 4 composites with 10% PVA and various BC concentrations (0, 0.15, 0.23, 0.31, and 0.61%) for cycle 6.

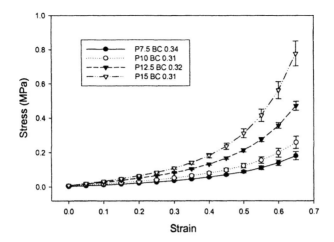

Figure 6. Stress-strain curves of 4 composites with — 0.3% bacterial cellulose and different PVA concentrations (7.5,]0,]2.5,]5)for cycle 6.

composites at a strain of 65%, all the treatment groups were found to be statistically different (P<0.05).

One important observation is that the mechanical properties can be controlled over a very broad range, so that any tissue with mechanical properties falling within this range of stress-strain curves can be matched by altering and controlling a combination of variables, including PVA and BC concentrations, number of freeze/thaw cycles, thawing rate, and freezing holding time.

The PVA-BC interactions are probably based on two mechanisms: a high degree of hydrogen bonding and BC crystallites serving as nuclei for cross-linking of PVA polymer molecules. During the freeze/thaw cycles, ice crystals in the amorphous regions force the polymer chains into regions of high local polymer concentration, where they align to form crystallites and polymer chain entanglements (74, 75). Willcox et al. (76) studied the PVA microstructure and concluded that in the initial cycle a few percent of the chain segments crystallize into 3-8 nm junctions separated by amorphous regions of around 30 nm. Further cycling augments the level of crystallinity, transforming the microstructure into a fibrillar network with ~30 nm pores. Also, the concentration of polymer (both PVA and BC) in solution affects the modulus because the higher the initial concentration, the greater the number of polymer chains in solution available to create crystalline regions in the cross-linked PVA nanocomposite.

BC is a highly hydrophilic crystalline nanofiber, with high mechanical strength (56). These BC crystallites could alsb serve as sites for further arrangement of the PVA chains, thus favoring the interaction and formation of PVA crystallites as the thermal cycling increases, since additional crystallites can form around the BC nanofibers. The strong bonding between the BC fibers and the PVA matrix allows the matrix to successfully transfer the load to the

236

fibers. The PVA matrix performs a function similar to elastin, allowing the material to stretch without a large increase in tension, while the BC fibers behave similarly to collagen, the load bearing component, contributing to the material tension as the sample is progressively stretched (*61, 77*). These effects can be seen in Figures 5 and 6, where the nanocomposite curves display higher stiffness.

To demonstrate the ability to tune these composites to match the properties of specific tissues, we compare their mechanical properties with those of aortic heart valve leaflets (Figure 7) and porcine aorta (Figure 8). The tensile properties of aortic heart valve leaflets were not perfectly matched in the circumferential direction, but we were able to get close. On the other hand, porcine aorta was closely matched in either direction by at least one type of PVA-BC nanocomposite.

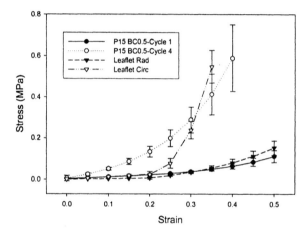

Figure 7. Stress-strain curves for the cycles 1 and 4 of the 15% PVA with 0.5% BC composite and the circumferential and radial curves for aortic heart valve.

In Figure 7, it can be seen that the modulus of the heart valve tissue at low strains (<20%) is very close to zero, indicating almost no resistance to stretching. There is no clear match of the heart valve leaflet in the circumferential direction, but a comparison was made with the 15% PVA with 0.5% cycle 4. However it is clearly seen that the heart valve tissue stiffness drastically increases after a strain of about 20%. ANOVA was applied to the stress of both directions of aortic heart valves and the 15% PVA with 0.5% BC (cycles 4 and 1) at a strain of 30%. All the treatment groups were found to be statistically different (P<0.05), except for the aortic valve in the radial direction and the PVA-BC nanocomposite (cycle 1).

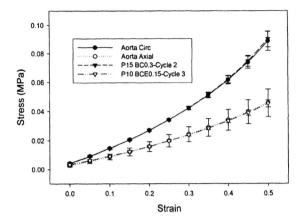

Figure 8. Stress-strain curves for 15% PVA with 0.3% bacterial cellulose (cycles 2) and the 10% PVA with 0.15% bacterial cellulose (cycle 3) showing a close match with porcine aorta in circumferential and axial directions.

In Figure 8, when ANOVA was applied to the stress-strain curves in both directions of porcine aortic to compare with the 15% PVA with 0.3% bacterial cellulose (cycles 2) and the 10% PVA with 0.15% bacterial cellulose (cycle 3) at a strain of 30%, no statistical difference was found (P>0.05). From these results we conclude that a PVA-BC nanocomposite with a mechanical response and elastic modulus closely matching aorta in either a circumferential or axial direction can be produced. Therefore, it was possible to have selective matching of properties of porcine aorta, but since PVA or PVA-BC nanocomposites display isotropic behavior, none of the compositions can simultaneously match properties in both the circumferential and axial directions.

In summary, the PVA-BC nanocomposite developed here is a biomaterial with mechanical properties tunable over a broad range. Thus, depending on the particular application, PVA-BC nanocomposites with mechanical properties closely approximating those of many tissues can be prepared.

References

1. Geyer, U., Heinze, T., Stein, A., Klemm, D., Marsch, S., Schumann, D., Schmauder, H. P., *International Journal of Biological Molecules,* **1994**, *16*: 343-347.
2. Fontana, J. D., A. M. de Souza, C. K. Fontana, I. L. Torriani, J. C. Moreschi, B. J. Gallotti, S. J. de Souza, G. P. Narcisco, J. A. Bichara, and L. F. X Farah, *Journal of Applied Biochemistry and Biotechnology,* **1990**, *25/25*, 253-264.

238

3. Krieger, J. *Chemical and Engineering News,* March 2 **1990**, *68*, 37-37.
4. Williams, W.S. and R. E. Cannon, *Applied and Environmental Microbiology*, **1989**, *55*, 2448-2452.
5. Cook, K. E., Colvin, J. R., *Current Microbiology,* **1980**, *3*, 203-205.
6. Schramm, M. and S. Hestrin, *Journal of General Microbiology,* **1954**, *11*, 123-129.
7. Williams, W.S. and R. E. Cannon, *Applied and Environmental Microbiology*, **1989**, *55*, 2448-2452
8. Brown, A. J. *Journal of the Chemical Society of London,* **1886**, *49*, 432-439.
9. Iguchi, M., Yamanaka, S., and Budhiono, A., *Journal of Material Science,* **2000**, *35*, 261-270.
10. Yoshino, K., Matsuoka, R., Nogami, K., Arai, H., Yamanaka, S., Watanabe, K., Takahashi, M., and Honda, M., *Synthetic Metals,* **1991**, *41-43*, 1593-1596.
11. Shibazaki, H., Kuga, S., Onabe, F., Usuda, M., *Journal of Applied Polymer Science,* **1993**, *50*, 965-969.
12. Byrom, D., "Microbial Cellulose", Biomaterials: Novel Materials from Biological Sources, ed. D. Byrom, Macmillan Publishers Ltd.: UK **1991**.
13. Klemm, D., Schumann, D., Udhardt, U., Marsch, S. *Progress in Polymer Science,* **2001**, *26*, 1561-1603.
14. Svensson, A., Nicklasson, E., Harrah, T., Panilaitis, B., Kaplan, D. L., Brittberg, M., Gatenholm, P. *Biomaterials,* **2005**, *26*, 419-431.
15. Muhlethaler, K., *Biochimica et Biophysica Acta,* **1949**, *3*, 527-535.
16. Ohad, I., Danon M. D., and Hestrin, S., *Journal of Cell Biology,* **1962**, *12*, 31-46.
17. Colvin, J., *Journal of Cell Biology,* **1963**, *17*, 105-109.
18. Yamanaka, S., Watanabe, K., Kitamura, N., Iguchi, M., Mitsuhashi, S., Nishi Y., and Uryu, M., *Journal of Material Science,* **1989**, *24*, 3141-3145.
19. Marx-Figini, M. "The control of Molecular Weight and Molecular Weight Distribution in the Biogenesis of Cellulose", Cellulose and Other Natural Polymer Systems ed. R. M. Brown Jr. Plenum Press: New York **1982**.
20. Marx-Figini, M. and Pion, B. G., *Biochimica et Biophysica Acta,* **1974**, *338*, 382-393.
21. Tahara, N., Tabuchi, M., Kunihiko, W., Yano, H., Morinaga, Y., Yoshinaga, F., *Bioscience, Biotechnology and Biochemistry,* **1997**, *61*, 1862-1865.
22. Tonouchi, N., Tahara, T., Tsuchida, T., Yoshinaga, F., Beppu, T., and Horinouchi, S., *Bioscience, Biotechnology and Biochemistry,* **1995**, *59*, 805-808.
23. Watanabe, K., Tabuchi, M., Ishikawa, A., Takemura, H., Tsuchida, T., Morinaga Y., and Yoshinaga, F., *Bioscience, Biotechnology, and Biochemistry,* **1998**, *62*, 1290-1292.
24. Hestrin, S., Aschner, M., and Marger, J., *Nature,* **1947**, *159*, 64-65.
25. Schramm, M., Gromet, Z., and Hestrin, S., *Biochemical Journal,* **1957**, *67*, 669-679.

239

26. Borzani, W. and de Souza, J., *Biotechnology Letters*, **1995**, *17*, 1271-1272.
27. Schramm, M. and Hestrin, S., *Journal of General Microbiology*, **1954**, *11*, 123-129.
28. Dudman, W. F., *Journal of General Microbiology*, **1960**, *22*, 25-39.
29. Watanabe, K. and Yamanaka, S., *Bioscience, Biotechnology and Biochemistry*, **1995**, *59*, 65-68.
30. Verschuren, P. G., Cardona, T. D., Nout, M. J. R., DeGooijer, D., van den Heuvel, J. C., *Journal of Bioscience and Bioengineering*, **2000**, *89*, 414-419.
31. Hestrin, S., and Schramm, M., *Biochemistry Journal*, 58: 345-352 (1954)
32. Toyosaki, H, Kojima, Y., Tsuchida, T., Hoshino, K., Yamada, Y., and F. Yoshinaga, *Journal of General and Applied Microbiology*, **1995**, *41*, 307-314.
33. Sakairi, N., Asano, H., Ogawa, M., Nishi, N., and Tokura, S., *Carbohydrate Polymers*, **1998**, *35*, 233-237.
34. Kojima, Y., Seto, A., Tonouchi, N., Tsuchida, T., and Yoshinaga, F. *Bioscience, Biotechnology, and Biochemistry*, **1997**, *61*, 1585-1586.
35. Toyosaki, H., Naritomi, T., Seto, A., Matsuoka, M., Tsuchida, T. and Yoshinaga, F., *Bioscience, Biotechnology and Biochemistry*, **1995**, *59*, 1498-1502.
36. Matsouka, M., T. Tshuchida, K. Matsushita, O. Adachi and F. Yoshinaga, *Bioscience, Biotechnology, and Biochemistry*, **1996**, *60*, 575-579.
37. Bungay, III H. R., G. C. Serafica, **1999**, US Patent Number 5,955,926.
38. Tsuchida, T. and F. Yoshinaga, *Pure and Applied Chemistry*, **1997**, *69*, 2453-2458.
39. Ougiya, H., K. Watanabe, T. Matsumura and F. Yoshinaga, *Bioscience Biotechnology and Biochemistry*, **1998**, *62*, 1714-1719.
40. Byrom, D., **1990**, US Patent Number 4,929,550.
41. Krystynowicz, A., Czaja, W., Wiktorowska-Jezierska, A., Goncalves-Miskiewicz, M., Turkiewicz, M., Bielecki, S. *Journal of Industrial Microbiology & Biotechnology*, **2002**, *29*, 189-195.
42. Johnson, D. C., and A. N. Neogi, **1989**, US Patent Number 4,863,565.
43. Ben-Bassat, A., K. D. Coddington, and D. C. Johnson, **1992**, US Patent number 5,114,849.
44. Moo-Young, M., T Hirose, K. H. Geiger, *Biotechnology and Bioengineering*, **1969**, *11*, 393-412.
45. Joseph, G., Rowe, G. E., Margaritis, A., Wan, W. *Journal of Chemical Technology and Biotechnology*, **2003**, *78*, 964-970.
46. Ishida, Takehiko., Mitarai, Makoto., Sugano, Yasushi., Shoda, Makoto. *Biotechnology and Bioengineering*, **2003**, *83*, 474-478.
47. Son, Hong-Joo., Heo, Moon-Su., Kim, Yong-Gyun., Lee, Sang-Joon. *Biotechnology and Applied Biochemistry* **2001**, *33*, 1-5.
48. Seto, A., Y. Kojima, N. Tonouchi, T. Tsuchida and F. Yoshinaga, *Bioscience, Biotechnology and Bioengineering*, **1997**, *61*, 735-367.

49. Gromet-Elhanan, Z. and S. Hestrin, *Journal of Bacteriology,* **1963**, *85,* 284-292.
50. Matsouka, M., T. Tshuchida, K. Matsushita, O. Adachi and F. Yoshinaga, *Bioscience, Biotechnology, and Biochemistry,* **1996**, *60,* 575-579.
51. Naritomi, T., T. Kouda, H. Yano and F. Yoshinaga, *Journal of Fermentation and Bioengineering,* **1998**, *85,* 89-95.
52. Zabriskie, D. W., Trader's Guide to Fermentation Media Formulation, Traders Protein: Memphis, Tenn., **1999**.
53. Yagi H. and F. Yoshida *Industrial and Engineering Chemistry Process Design and Development,* **1975**, *14,* 488-493.
54. Eichhorn, S. J. and Young, R. J. *Cellulose,* **2001**, *8,* 197-207.
55. Horii, F., Hirai, A., Kitamaru, R. *Polymer Bulletin,* **1982**, *8,* 162-170.
56. Guhados, G., Wan, W.K., Hutter, J.L., *Langmuir,* **2005**, *21,* 6642-6646.
57. Hutter, J. L., Bechhoefer, J., *Rev Sci Instrum.,* 64: 1868-1873, (1993) with corrections pointed out by Butt, H.-J., Jaschke, M. *Nanotech.,* 6:1–7 (1995) and Walters, D. A., Cleveland, J. P., Thomson, N. H., Hansma, P. K., Wendman, M. A., Gurley, C., Elings, V. *Rev Sci Instrum.,* **1996**, *67,* 3580–3590.
58. Cleveland, J. P., Manne, S., Bocek, D., Hansma, P. K. *Rev Sci Instrum.,* 64: 403-405 (1993), with corrections pointed out by Sader, J.E., Larson, I., Mulvaney, P., White, L.R. *Rev Sci Instrum.,* **1995**, *66:* 3789-3798.
59. Timoshenko, S. and Gere, J.M. Mechanics of Materials, 4th ed.; PWS Publishing Co.: Boston, MA, **1997**.
60. Wan W.K., Campbell G, Zhang Z.F., Hui A.J., Boughner D.R. *J Biomed Mater Res.,* **2002**, *63,* 854-861.
61. Fung Y.C., Biomechanics: Mechanical Properties of Living Tissue. 2nd ed. Springer-Verlag, New York **1993**.
62. Mayne A.S., Christie C.W., Smaill B.H., Hunter P.J., Barratt-Boyes B.C., *J Thorac Cardiovasc Surg,* **1989**, *98,* 170-180.
63. Leeson-Dietrich J., Boughner D., Vesely I. *J Heart Valve Dis,* **1995**, *4,* 88-94.
64. Salvetat, J.-P., Briggs, G. A. D., Bonard, J.-M., Bacsa, R. R., Kulik, A. J., Stockli, T., Burnham, N. A., Forro, L., *Physical Review Letters,* **1999**, *82,* 944-947.
65. Koyama, M., Helbert, W., Imai, T., Sugiyama, J., Henrissat, B. *Proceedings of the National Academy of Sciences (USA)* **1997**, *94,* 9091-9095.
66. Xue, L., Greisler, H.P., *J Vasc Surg.,* **2003**, *37,* 472-478.
67. Peppas, N.A., Hassan, C.M., *Adv Polym Sci.,* **2000**, *8,* 37-41.
68. Hassan, C.M., Peppas, N.A., *Macromolecules,* **2000**, *33:* 2472-2479.
69. Hayash, K., Stergiopulos, N., Meister, J., Greenwald, S.E., Rachev, A., Techniques in the determination of the mechanical properties and constitutive laws of arterial walls. In: Leondes C, editor. Biomechanical Techniques: Techniques and Applications – Volume 2. Boca Raton, FL: CRC Press, **2001**.

70. Stauffer, S.R., Peppas, N.A., *Polymer,* **1992**, *33*, 3932-3935.
71. Gordon, M.J., Controlling the Mechanical Properties of PVA Hydrogels For Biomedical Applications. MESc Thesis, University of Western Ontario, **1999**.
72. Schoen, F.J., Levy, R.J., Founder's award, 25th annual meeting of the society for biomaterials, perspectives, providence, RI, April 28-May 2, 1999. Tissue heart valves: current challenges and future research perspectives. *J Biomed Mater Res.,* **1999**, *47*, 439-65.
73. Chu, K.C., Rutt, B.K., *Magnetic Reson Med.* **1997**, *37*, 314-319.
74. Mori, Y., Tokura, H., Yoshikawa, M., *J Mater Sci.,* **1997**, *32*, 491-496.
75. Lozinsky, V.I., Plieva, F.M., *Enzyme Microb Technol.* **1998**, *23*, 227-242.
76. Willcox, P.J., Howie, D.W., Schmidt-Rohr K., Hoagland, D.A., Gido, S.P., Pudjijanto, S., *J Polym Sci [B],* **1999**, *37*, 3438-3454.
77. Ratner B.D., Hoffman A.S., Schoen F.J., Lemons J.E., editors. Biomaterials Science: An Introduction to Materials in Medicine. 2nd ed. San Diego: Elsevier Academic Press; **2004**.

Indexes

Author Index

Subject Index